The Secret of Apollo

New Series in NASA History

Roger D. Launius

SERIES EDITOR

Before Lift-off: The Making of a Space Shuttle Crew

HENRY S. F. COOPER, JR.

The Space Station Decision: Incremental Politics and Technological Choice

HOWARD E. MCCURDY

Exploring the Sun: Solar Science since Galileo

KARL HUFBAUER

Inside NASA: High Technology and Organizational Change in the U.S. Space Program

HOWARD E. MCCURDY

Powering Apollo: James E. Webb of NASA

W. HENRY LAMBRIGHT

NASA and the Space Industry

JOAN LISA BROMBERG

Taking Science to the Moon: Lunar Experiments and the Apollo Program

DONALD A. BEATTIE

Faster, Better, Cheaper: Low-Cost Innovation in the U.S. Space Program

HOWARD E. MCCURDY

The Secret of Apollo: Systems Management in American and European Space Programs

STEPHEN B. JOHNSON

Space Policy in the Twenty-First Century

EDITED BY W. HENRY LAMBRIGHT

THE SECRET OF APOLLO

Systems Management in American and European Space Programs

Stephen B. Johnson

The Johns Hopkins University Press
Baltimore and London

Printed in the United States of America on acid-free paper

Johns Hopkins Paperbacks edition, 2006
2 4 6 8 7 5 3 1

The Johns Hopkins University Press
2715 North Charles Street
Baltimore, Maryland 21218-4363
www.press.jhu.edu

The Library of Congress has cataloged the hardcover edition of this book as follows:

Johnson, Stephen B., 1959–
The secret of Apollo : systems management in American and European space programs /
Stephen B. Johnson.
p. cm. — (New series in NASA history)
Includes bibliographical references and index.
ISBN 0-8018-6898-X (hardcover : alk. paper)
1. Astronautics, Military—United States—Management.
2. Astronautics—United States—Management. 3. Astronautics, Military—Europe—Management. 4. Astronautics—Europe—Management. I. Title. II. Series.
UG1523 .J645 2002 629.4′0973—dc21
2001005688

ISBN 0-8018-8542-6 (pbk. : alk. paper)

A catalog record for this book is available from the British Library.

To Diane

Contents

Illustrations

Preface and Acknowledgments

This book builds on historical research I carried out over the last seven years and also on my own history and values. I did not begin with the intention of studying systems management or systems engineering, subjects familiar to me from my background in the aerospace industry. In fact, I made some effort at the start *not* to do so, to avoid my own biases. Originally, I wanted to use my aerospace experience but also to separate myself somewhat from it so as to look at the history of the aerospace industry from a more detached standpoint. I eventually decided to investigate more closely the Spacelab program, a joint effort of the National Aeronautics and Space Administration (NASA) and the European Space Agency (ESA). This seemed a good choice because I knew something of space technology but little about manned laboratories or ESA.

Spacelab looked like a good case of technology transfer from the United States to Europe. Yet I found little novelty in *Spacelab*'s hardware technology, and neither did the Europeans. So why were they interested in this project? They wanted to learn how to manage the development of large, complex space systems—that is, the methods of "systems management." Soon I encountered the "technology gap" and "management gap" literature, the pervasive rhetoric about "systems," and the belief in the Apollo program as a model for how to solve social as well as technical problems. This was a worthy topic, particularly because no other historian had investigated it.

Systems approaches emphasize integrative features and the elements of human cooperation necessary to organize complex activities and technologies. Believing that humans are irrational, I find the creation of huge, orderly, rational technologies almost miraculous. I had never pondered the deeper implications of cooperative efforts amid irrationality and conflict, and this project has enabled me to do so.

I owe a debt of thanks to many. At the History Office at NASA headquarters, Roger Launius, Lee Saegesser, and Colin Fries were helpful in guiding me through the collections. Julie Reiz, Elizabeth Moorthy, and Michael Hooks provided excellent service at the Jet Propulsion Laboratory archives, declassifying numerous documents for my rather diffuse research. At the European Community archives at the European University Institute (EUI) in Florence, Italy, Gherardo Bonini located numerous documents and provided many records I would not have otherwise noticed, sending some to me later when I found that I needed more information. The Technical Information and Documentation Center at the ESA's European Space Technology Centre (ESTEC) opened its doors (literally) for me, allowing me to rummage through storerooms full of documents, as well as its collection of historical materials. ESTEC's Lilian Viviani, Lhorens Marie, Sarah Humphrey, and director Jean-Jacques Regnier were all extremely helpful. John Krige, who headed the European Space History project, provided travel funding to visit the EUI and ESTEC archives. I am particularly grateful for his help and trust in me, because he jump-started my research when it was in its very early stages.

In 1998 and 1999 I performed related research for the Air Force History Support Office, contract number F4964298P0148. This provided travel funds and support for my graduate student Phil Smith. I am grateful to Phil for doing much of the "legwork" to dig up archival materials in the Boston area and at Maxwell Air Force Base in Montgomery, Alabama. Chuck Wood in the Space Studies Department of the University of North Dakota encouraged me in this work, and I appreciate his understanding and support for this research among my other faculty duties. I thank Cargill Hall, Rich Davis, and Priscilla Jones for their efforts on my behalf in the History Support Office. Harry Waldron at the Space and Missile Center was extremely helpful in gathering further materials on ballistic missiles.

Also providing funding for my research was the University of Minnesota Research and Teaching Grant and Dissertation Fellowship program. The professors at the University of Minnesota with whom I studied from 1992 to 1997 taught me much of what it means to be a historian. David Good and George

Green introduced me to the literature of economic and business history. Ron Giere inspired me to consider philosophical and cognitive issues and to recognize the value of theory, not just for philosophy, but for history as well. Ed Layton and Alan Shapiro stressed the importance of thorough research. Roger Stuewer's kindness and concern brought me to Minnesota to begin with, and his courses in the history of nuclear physics were important for my understanding of the European background of large-scale technology development. Robert Seidel helped me to write with more conciseness and clarity and to see several implicit assumptions that I had made. My adviser, Arthur Norberg, prodded me to keep moving and to maintain a steady focus on the core issues—the concerns that led me to this project. He kept bringing the "big picture" questions to my attention.

A few scholars significantly influenced my thinking. Joanne Yates's approach in *Control Through Communication* formed an important early model for my work. James Beniger, Ross Thomson, Theodore Porter, Tom Hughes, and Daniel Nelson all influenced this research as well. John Lonnquest and Glenn Bugos performed recent research on the air force and navy that directly links to mine.

A number of scholars have reviewed this manuscript, either as a whole or in articles derived from it, and given me significant feedback that has helped in various ways. These include Alex Roland, Harvey Sapolsky, John Krige, John Staudenmaier, Roger Launius, Tom Hughes, R. Cargill Hall, John Lonnquest, my committee at the University of Minnesota, and the anonymous reviewers with *Technology and Culture, History and Technology, History of Technology, Air Power History,* the Air Force History Support Office, and the Johns Hopkins University Press. The anonymous Johns Hopkins University Press reviewer gave me excellent critiques. I owe to him or her the insight that concurrency is not really a management method but rather a strategy that requires a strong management method to succeed.

To the extent that this work succeeds, I owe all of these people who helped me along the way. Any flaws that remain are my own.

Finally, I must thank my wife, Diane, and my two sons, Casey and Travis, for being patient with me through this long and arduous process. Only as I look back now do I realize how difficult it has been.

I sincerely hope that this work helps others recognize that the "systems" in which we all take part are our own creations. They help or hinder us, depending upon our individual and collective goals. Regardless of our feelings about them, they are among the pervasive bonds that hold our society together.

Abbreviations and Acronyms

AAS	American Astronautical Society
ABMA	Army Ballistic Missile Agency
AF	air force
AFHRA	Air Force Historical Research Agency
AFSC	Air Force Systems Command
AMC	Air Materiel Command
ARDC	Air Research and Development Command
ASAT	Arbeitsgemeinschaft Satelitenträger
AT&T	American Telephone and Telegraph
ATC	Air Training Command
BMC	Ballistic Missile Command
BSD	Ballistic Systems Division
CBI	Charles Babbage Institute
CCB	configuration control board
CERN	Conseil Européen pour Recherche Nucléaire
CETS	Conférence Europééne de Télécommunications par Satellites
CNES	Centre National d'Études Spatiales
COPERS	Commission Préparatoire Europééne de Recherches Spatiales
CPFF	cost plus fixed fee
DCAS	Deputy Commander for Aerospace Systems
DOD	Department of Defense
ELDO	European Space Vehicle Launcher Development Organisation
ERNO	Entwicklungsring Nord
ESA	European Space Agency
ESRO	European Space Research Organisation
ESTEC	European Space Technology Centre

EUI	European University Institute
GE	General Electric
GSFC	Goddard Space Flight Center
HAC	Hughes Aircraft Company
HAEUI	Historical Archives of the European University Institute
HEOS	Highly Eccentric Orbit Satellite
HQ	headquarters
HSO	History Support Office
IBM	International Business Machines
ICBM	intercontinental ballistic missile
IRBM	intermediate-range ballistic missile
JPL	Jet Propulsion Laboratory
JPLA	JPL Archives
JSC	Johnson Space Center
KSC	Kennedy Space Center
LC/SPP	Library of Congress, Samuel Phillips Papers
LCT	Laboratoire Central de Télécommunications
LEM	Lunar Excursion Module
MAU	million accounting units
MBB	Messerschmitt-Bölkow-Blohm
MESH	Matra, Entwicklungsring Nord, Saab, and Hawker Siddeley
MIS	management information system
MIT	Massachusetts Institute of Technology
MSC	Manned Spacecraft Center
MSFC	Marshall Space Flight Center
NAA	North American Aviation
NACA	National Advisory Committee for Aeronautics
NASA	National Aeronautics and Space Administration
NASAHO	NASA History Office
NATO	North Atlantic Treaty Organization
OECD	Organization for Economic Cooperation and Development
OMSF	Office of Manned Space Flight
PDR	Preliminary Design Review
PERT	Program Evaluation and Review Technique (or Research Task)
QA	quality assurance

R&D	research and development
RCA	Radio Corporation of America
RFP	request for proposal
R-W	Ramo-Wooldridge
S&ID	Space and Information Systems Division
SAC	Strategic Air Command
SAPO	Special Aircraft Projects Office
SE	systems engineering
SEREB	Société pour l'Étude et la Réalisation d'Engins Balistiques
SETIS	Société d'Étude et d'Intégration de Systèmes Spatiaux
SMC	USAF Space and Missile Center
STG	Space Task Group
STL	Space Technology Laboratories
TD	Thor-Delta
TQM	total quality management
TRW	Thompson-Ramo-Wooldridge
USAF	U.S. Air Force
WADC	Wright Air Development Center
WDD	Western Development Division
WSPO	Weapon System Project Office

The Secret of Apollo

Management and the Control of Research and Development

Control . . . depends upon information and activities involving information: information processing, programming, decision, and communication.

—James Beniger, *The Control Revolution,* 1986

Since at least the Middle Ages, Western society's fascination with sophisticated technology has demanded organizational solutions. By the middle of the nineteenth century, railroads in Europe and the United States required professional managers to run them.[1] As the scale of operations increased, executives developed "systematic management" to coordinate and control their midlevel personnel.[2] At the beginning of the twentieth century, Frederick Winslow Taylor, publishing his major work in 1911, devised a means—by way of "scientific management"—of extending managerial influence to the factory floors of increasingly large industrial enterprises.[3] In both systematic and scientific management, information provided the levers that managers used to control their subordinates. Frequently working with engineers, managers gathered information from lower-level staff and then used that knowledge to reorganize work processes and control employees.[4]

Scientists and engineers eventually posed far more difficult challenges to managers. Universities trained these "knowledge workers," as management consultant Peter Drucker referred to them in the late 1940s, to be dedicated to their careers and their specialties, not to their employers. They generated new ideas in an undefined process that no one could routinize, thus ruling out scientific management techniques. Their specialized knowledge placed them beyond the competence of most managers. Even if technical personnel wanted to share their knowledge with managers (which they typically did not), they could not clearly describe their creative process. Only after the fact, it seemed, could managers control the products or the technologists who created them. Even so, managers seldom perceived research and development (R&D) management as a critical issue.[5] Drucker suggested a solution he called management by objectives. According to this approach, managers and professionals jointly negotiated the objectives for the agency or firm on the one hand and for the individuals on the other, each worker agreeing to the terms. Individuals and agencies or firms would harmonize their respective goals.[6]

The management-by-objectives strategy worked reasonably well for managers overseeing individual knowledge workers, but it did little to coordinate the efforts of scientists and engineers on large projects, on which experts organized (or disagreed) along disciplinary lines and could form only temporary committees for the exchange of information. Much like with the unique and short-lived Manhattan Project, the experience of complicated programs such as ballistic missiles demonstrated that traditional organizational schemes would not suffice. Scientists and engineers found that they needed some individuals to coordinate the information flowing among working groups. These "systems engineers" created and maintained documents that reflected the current design, and they coordinated design changes with all those involved in the program. Perceptive managers and military officers realized that central design coordination allowed them to gain control of both the creative process and its lively if unruly knowledge workers.

This study examines how scientists and engineers created a process to coordinate large-scale technology development—systems management—and how managers and military officers modified and gained control of it. The story owes a debt to the insights of Max Weber, who noted long ago that modern organizations form standardized rules and procedures that create and sustain bureaucracies.[7] Scholars since then have elaborated upon the development of these procedures as a process of "knowledge codification," one that can be formally internal to individuals or informally contained in the communications between or among individuals.[8] For organizations to learn, to adapt, and to sustain adaptations, they must have processes that are both flexible and durable. Recent scholarship on these so-called learning organizations has pursued and elaborated on this view, providing a perspective congenial to a historical analysis of management. By means of communication, feedback, and codification, organizations can be said to learn and retain knowledge.[9]

Systems management first developed in the air defense and ballistic missile programs of the 1950s, across many aerospace organizations. These programs, like any other large-scale technologies, came into being as a result of negotiations among various organizations, classes, and interest groups.[10] Scientists typically created the core ideas behind new systems or the critical elements that made them possible or useful. Engineers developed the subsystems and integrated them into a complex vehicle. Military officers promoted these complex vehicles as a means of besting their Cold War foes. Managers controlled the resources required to produce the new systems. Systems management was embraced because it assigned each of these groups a standard role in the technology development process. Systems management became the core process of aerospace R&D institutions, modeled largely on management techniques developed on army and air force ballistic missile programs. Methods developed for air defense systems paralleled those for ballistic missiles, but in the bureaucratic battles of the early 1960s, ballistic missile officers and their methods triumphed, forming the basis for the air force's procurement regulations.[11]

This book thus traces a path through the literature on the history and politics of aerospace development and weapons procurement.[12] Instead of providing another case study of a particular project or organization, it pieces together a story from elements that include military and civilian organizations in the United States and Europe. This approach has the distinct advantage of providing cross-organizational and cross-cultural perspectives on the subject, as well as showing the dynamics of the transfer of management methods. NASA and the European programs encountered the same kinds of technical and social issues that the air force and the Jet Propulsion Laboratory (JPL) had previously come upon, and ultimately they looked outside of their organizations to help resolve the problems. NASA looked to the air force (and to a lesser degree to JPL), and a few years later the Europeans gleaned their methods from NASA. The Apollo program became a highly visible icon of American managerial skill—the symbol of the difference between American technical prowess and European technical retardation in the 1960s and early 1970s.

European frustration reached its peak in 1969, when NASA put men on the Moon while the European Space Vehicle Launcher Development Organisation (ELDO) endured yet another failure of its launcher. ELDO only haphazardly adopted American management methods, and the lack of authority meant that those that ELDO did adopt could not be consistently implemented. The failures of ELDO ultimately proved to be the spur for the Europeans to overcome their historic hostilities and create a highly successful integrated space organization, the European Space Agency. This new agency and its predecessor, the European Space Research Organisation, borrowed extensively from NASA and its contractors. NASA's management methods, when adapted to the European environment, became key ingredients in Europe's subsequent successful space program. The air force, the army's (and later NASA's) JPL, NASA's manned space programs, and the European integrated space programs all learned that spending more to ensure success was less expensive than failure.

The modern aerospace industry is paradoxical. It is both innovative, as its various air and space products attest, and bureaucratic, as evidenced by the hundreds of engineers assigned to each project and the overpriced components used. How can these two characteristics coexist? The answer lies in the nature of aerospace products, which must be extraordinarily dependable and robust, and in the processes that the industry uses to ensure extraordinary dependability. Spacecraft that fail as they approach Mars cannot be repaired. Hundreds can lose their lives if an aircraft crashes. The media's dramatization of aerospace failures is itself an indication that these failures are not the norm. In a hotly contested Cold War race for technical superiority, the extreme environment of space exacted its toll in numerous failures of extremely expensive systems. Those funding the race demanded results. In response, development organizations created what few expected and even fewer wanted—a bureaucracy for innovation. To begin to understand this apparent contradiction in terms, we must first understand the exacting nature of space technologies and the concerns of those who create them.

ONE

Social and Technical Issues of Spaceflight

> Europe's lag seems to concern *methods of organization* above all. The Americans know how to work in our countries better than we do ourselves. This is not a matter of "brain power" in the traditional sense of the term, but of organization, education, and training.
>
> —Jean-Jacques Servan-Schreiber, 1967

July 1969 marked two events in humanity's exploration of space. One became an international symbol of technological prowess; the other, a mere historical footnote, another dismal failure of a hapless organization.

"One small step for man, one giant leap for mankind." These words of American astronaut Neil Armstrong, spoken as he stepped onto the surface of the Moon in July 1969, represented the views not only of the National Aeronautics and Space Administration (NASA) but also of numerous Americans and space enthusiasts around the world. Many journalists, government heads, and industrial leaders believed that the Apollo program responsible for Armstrong's exotic walk had been a tremendous success. They marveled at NASA's ability to organize and direct hundreds of organizations and hundreds of thousands of individuals toward a single end. Even Congress was impressed, holding hearings to uncover the managerial secrets of NASA's success.[1]

Apollo was the centerpiece of NASA's efforts in the 1960s—the United States' most prestigious entry in the propaganda war with the Soviet Union. Purportedly, the massive program cost more than $19 billion through the first Moon landing and used 300,000 individuals working for 20,000 contractors and 200 universities in 80 countries.[2] It was a visual, technological, and publicity tour-de-force, capturing the world's attention with television broadcasts of the *Apollo 8* voyage to the Moon during Christmas 1968, the *Apollo 11* landing, and the dramatic near-disaster of *Apollo 13* in April 1970. Whatever else might be said about the program, it was an impressive technological feat.

This American achievement looked all the more impressive to European observers, who on July 3, 1969, witnessed the fourth consecutive failure of their own rocket, the grandiosely named *Europa I*. Whereas *Apollo*'s mandate included a presidential directive, national pride, and an all-out competition with the Soviet Union, *Europa I* began as a cast-off ballistic missile searching for a mission. When British leaders decided to use American missile technology in the late 1950s, their own obsolete rocket, *Blue Streak*, became expendable. The British decided to market it as the first stage of a European rocket, simultaneously salvaging their investment and signaling British willingness to cooperate with France, a gesture they hoped would lead to British acceptance into the Common Market. Complex negotiations ensued, as first Britain and France—and then West Germany, Italy, Belgium, and the Netherlands—warily decided to build a European rocket. All the countries hoped to gain access to their neighbors' technologies and markets, while protecting their own as much as possible.

The European Space Vehicle Launcher Development Organisation (ELDO) reflected these national ambitions. Without the ability to let contracts or to direct the technical efforts, ELDO's Secretariat tried with growing dismay to integrate the vehicle, while its member states minimized access to the data necessary for such integration. Not surprisingly, costs rose precipitously and schedules slipped. After successful tests of the relatively mature British stage, every flight that tried to integrate stages failed miserably. The contrast between European failure and American success in July 1969 could not have been more stark, with American astronauts returning to Earth to lead a round-the-world publicity tour, while European managers and engineers defended themselves from criticism as they analyzed yet another explosion. ELDO's record of failure continued for more than four years before frustrated European leaders dissolved the organization and started over.

Apollo was a grand symbol, arguably the largest development program ever undertaken. Many observers noted the massive size and "sheer competence" of the program and concluded that one of the major factors in Apollo's success was its management.[3] Learning the organizational secrets of Apollo and the American space program was a primary motivation for European government and industry involvement in space programs.[4]

French journalist Jean-Jacques Servan-Schreiber gave European fears of American domination a voice and a focus in his best-selling 1967 book, *The American Challenge.* Servan-Schreiber argued that the European problems were due to inadequacies in European educational methods and institutions as well as the inflexibility of European management and government. The availability of university education to the average American led to better management of technology development in commercial aircraft, space, and computers. Europeans needed to learn the dominant American model for managing and organizing aerospace projects: systems management.

European space organizations needed to create or learn new methods to successfully develop space technology. Wernher von Braun's rocket team in Nazi Germany confronted major technical problems in the 1930s and 1940s, requiring new kinds of organizational processes. In the 1950s, the army's Jet Propulsion Laboratory (JPL) and the air force—through its industrial contractors—developed progressively larger, more complex, and more powerful ballistic missiles. Both groups encountered obstacles that the application of more gadgetry could not overcome. Like von Braun's group, these groups found that changes in organization and management were crucial. NASA's manned program confronted similar issues in the 1960s, resulting in major organizational innovations borrowed from the air force. In each case, the unique technical problems of spaceflight posed difficulties requiring social solutions—changes in how people within organizations in design and manufacturing processes related to one another.

Technical Challenges in Missile and Space Projects

Missiles were developed from simple rocketry experimentation between World Wars I and II. Experimenters such as Robert Goddard and Frank Malina in the United States, von Braun in Germany, Robert Esnault-Pelterie in France, and Valentin Glushko in the Soviet Union found rocketry experimentation a dangerous business. All of them had their share of spectacular mishaps and explosions before achieving occasional success.[5]

The most obvious reason for the difficulty of rocketry was the extreme volatility of the fluid or solid propellants. Aside from the dangers of handling exotic and explosive materials such as liquid oxygen and hydrogen, alcohols, and kerosenes, the combustion of these materials had to be powerful and controlled. This meant that engineers had to channel the explosive power so that the heat and force neither burst nor melted the combustion chamber or nozzle. Rocket engineers learned to cool the walls of the combustion chamber and nozzle by maintaining a flow of the volatile liquids near the chamber and nozzle walls to carry off excess heat. They also enforced strict cleanliness in manufacturing, because impurities or particles could and did lodge in valves and pumps, with catastrophic results. Enforcement of rigid cleanliness standards and methods was one of many social solutions to the technical problems of rocketry.[6]

Engineers controlled the explosive force of the combustion through carefully designed liquid feed systems to smoothly deliver fuel. Instabilities in the fuel flow caused irregularities in the combustion, which often careened out of control, leading to explosions. Hydrodynamic instability could also ensue if the geometry of the combustion chamber or nozzle was inappropriate. Engineers learned through experimentation the proper sizes, shapes, and relationships of the nozzle throat, nozzle taper, and combustion chamber geometry. Because of the nonlinearity of hydrodynamic interactions, which implied that mathematical analyses were of little help, experimentation rather than theory determined the problems and solutions. For the *Saturn* rocket engines, von Braun's engineers went so far as to explode small bombs in the rocket exhaust to create hydrodynamic instabilities, to make sure that the engine design could recover from them.[7] For solid fuels, the shape of the solid determined the shape of the combustion chamber. Years of experimentation at JPL eventually led to a star configuration for solid fuels that provided steady fuel combustion and a clear path for exiting hot gases. Once engineers determined the proper engine geometry, rigid control of manufacturing became utterly critical. The smallest imperfection could and did lead to catastrophic failure. Again, social control in the form of inspections and testing was essential to ensuring manufacturing quality.

Rocket engines create severe structural vibrations. Aircraft designers recognized that propellers caused severe vibrations, but only at specific frequencies related to the propeller rotation rate. Jet engines posed similar problems, but at higher frequencies corresponding to the more rapid rotation of turbojet rotors. Rocket engines were much more problematic because their vibrations were large and occurred at a wide range of nearly random frequencies. The loss of fuel also changed a rocket's resonant frequencies, at which the structure bent most readily. This caused breakage of structural joints and the mechanical connections of electrical equipment, making it difficult to fly sensitive electrical equipment such as vacuum tubes, radio receivers, and guidance systems. Vibrations also occurred because of fuel sloshing in the emptying tanks and fuel lines. These "pogo" problems could be tested only in flight.

Vibration problems could not generally be solved through isolated technical fixes. Because vibration affected electrical equipment and mechanical connections throughout the entire vehicle, this problem often became one of the first so-called system issues—it transcended the realm of the structural engineer, the propulsion expert, or the electrical engineer alone. In the 1950s, vibration problems led to the development of the new discipline of reliability and to the enhancement of the older discipline of quality assurance, both of which crossed the traditional boundaries between engineering disciplines.[8]

Reliability and quality control required the creation or enhancement of social and technical methods. First, engineers placed stronger emphasis on the selection and testing of electronic components. Parts to be used in missiles had to pass more stringent tests than those used elsewhere, including vibration tests using the new vibration, or "shake," tables. Second, technicians assembled and fastened electronic and mechanical components to electronic boards and other components using rigorous soldering and fastening methods. This required specialized training and certification of manufacturing workers. Third, to ensure that manufacturing personnel followed these procedures, quality assurance personnel witnessed and documented all manufacturing actions. Military authorities gave quality assurance personnel independent reporting and communication channels to avoid possible pressures from contractors or government officials. Fourth, all components used in missiles and spacecraft had to be qualified for the space environment through a series of vibration, vacuum, and thermal tests. The quality of the materials used in flight components, and the processes used to create them, had to be tightly controlled as well. This entailed extensive documentation and verification of materials as well as of processes used by the component manufacturers. Organizations traced every part from manufacturing through flight.[9]

Only when engineers solved the vibration and environmental problems could they be certain the rocket's electronic equipment would send the signals necessary to determine how it was performing. Unlike aircraft, rockets were automated. Although automatic machinery had grown in importance since the eighteenth century, rockets took automation to another level. Pilots could fly aircraft because the dynamics of an aircraft moving through the air were slow enough that pilots could react sufficiently fast to correct deviations from the desired path and orientation of the aircraft. The same does not hold true for rockets. Combustion instabilities inside rocket engines occur in tens of milliseconds, and explosions within 100 to 500 milliseconds thereafter, leaving no time for pilot reaction. In addition, early rockets had far too little thrust to carry something as heavy as a human.

Because rockets and satellites were fully automated, and also because they went on a one-way trip, determining if a rocket worked correctly was (and is) problematic. Engineers developed sophisticated signaling equipment to send performance data to the ground. Assuming that this telemetry equipment survived the launch and vibration of the rocket, it sent sensor data to a ground receiving station that recorded it for later analysis. Collecting and processing these data was one of the first applications of analog and digital computing. Engineers used the data to determine if subsystems worked correctly, or more importantly, to determine what went wrong if they did not. The military's system for problem reporting depended upon pilots, but contractors and engineers would handle problem reporting for the new technologies—a significant social change. Whereas in the former system, the military tested and flew aircraft prototypes, for the new technologies contractors flew prototypes coming off an assembly line of missiles and the military merely witnessed the tests.[10]

Extensive use of radio signals caused more problems. Engineers used radio signals to send telemetry to ground stations and to send guidance and destruct signals from ground stations to rockets. They carefully designed the electronics and wiring so that electromagnetic waves from one wire did not interfere with other wires or radio signals. As engineers integrated numerous electronic packages, the interference of these signals occasionally caused failures. The analysis of "electromagnetic interference" became another systems specialty.[11]

Automation also included the advanced planning and programming of rocket operations known as sequencing. Rocket and satellite engineers developed automatic electrical or mechanical means to open and close propulsion valves as well as fire pyrotechnics to separate stages, release the vehicle from the ground equipment, and otherwise change rocket functions. These "sequencers" were usually specially designed mechanical or electromechanical devices, but they soon became candidates for the application of digital computers. A surprising number of rocket and satellite failures resulted from improper sequencing or sequencer failures. For example, rocket stage separation required precise synchronization of the electrical signals that fired the pyrotechnic charges with the signals that governed the fuel valves and pumps controlling propellant flow. Because engineers sometimes used engine turbopumps to generate electrical power, failure to synchronize the signals for separation and engine firing could lead to a loss of sequencer electrical power. This in turn could lead to a collision between the lower and upper stages, to an engine explosion or failure to ignite, or to no separation. The solution to sequencing problems involved close communication among a variety of design and operations groups to ensure that the intricate sequence of mechanical and electrical operations took place in the proper order.[12]

Because satellites traveled into space by riding on rockets, they shared some of the same problems as rockets, as well as having a few unique features. Satellites had to survive launch vehicle vibrations, so satellite designers applied strict selection and inspection of components, rigorous soldering methods, and extensive testing. Because of the great distances involved—particularly for planetary probes—satellites required very high performance radio equipment for telemetry and for commands sent from the ground.[13]

Thermal control posed unique problems for spacecraft, in part because of the temperature extremes in space, and in part because heat is difficult to dissipate in a vacuum. On Earth, designers explicitly or implicitly use air currents to cool hot components. Without air, spacecraft thermal design required conduction of heat through metals to large surfaces where the heat could radiate into space. Engineers soon designed large vacuum chambers to test thermal designs, which became another systems specialty.

Unlike the space thermal environment, which could be reproduced in a vacuum chamber, weightlessness could not be simulated by Earth-based equipment. The primary effect of zero gravity was to force strict standards of cleanliness in spacecraft manufacturing. On Earth, dust, fluids, and other contaminants eventually settle to the bottom of the spacecraft or into corners where air currents slow. In space, fluids and particles float freely and can damage electrical components. Early spacecraft did not usually have this problem because many of them were spin stabilized, meaning that engineers designed them to spin like a gyroscope to hold a fixed orientation. The spin caused particles to adhere to the outside wall of the interior of the spacecraft, just as they would on the ground where the spacecraft would have been spin tested.

Later spacecraft like JPL's *Ranger* series used three-axis stabilization whereby the spacecraft did not spin. These spacecraft, which used small rocket engines known as thrusters to hold a fixed orientation, were the first to encounter problems with floating debris. For example, the most likely cause of the *Ranger 3* failure was a floating metal particle that shorted out two adjacent wires. To protect against such events, engineers developed conformal coating to insulate exposed pins and connectors. Designers also separated electrically hot pins and wires so that floating particles could not connect them. Engineers also reduced the number of particles by developing clean rooms where technicians assembled and tested spacecraft.

Many problems occurred when engineers or technicians integrated components or subsystems, so engineers came to pay particular attention to these interconnections, which they called interfaces. Interfaces are the boundaries between components, whether mechanical, electrical, human, or "logical," as in the case of connections between software components. Problems between components at interfaces are often trivial, such as mismatched connectors or differing electrical impedance, resistance, or voltages. Mismatches between humans and machines are sometimes obvious, such as a door too high for a human to reach, or an emergency latch that takes too long to operate. Others are subtle, such as a display that has too many data or a console with distracting lights. Finally, operational sequences are interfaces of a sort. Machines can be (and often are) so complicated to operate that they are effectively unusable. Spacecraft, whether manned or unmanned, are complex machines that can be operated only by people with extensive training or by the engineers who built them. Greater complexity increases the potential for operator error. It is probably more accurate to classify operator errors as errors in design of the human-machine interface.[14]

Many technical failures can be attributed to interface problems. Simple problems are as likely to occur as complex ones. The first time the Germans and Italians connected their portions of the *Europa* rocket, the diameters of the connecting rings did not match. Between the British first stage and the French second stage, electrical sequencing at separation caused complex interactions between the electrical systems on each stage, leading ultimately to failure. Other interface problems were subtle. Such was the failure of *Ranger 6* as it neared the Moon, ultimately traced to flash combustion of propellant outside of the first stage of the launch vehicle, which shorted out some poorly encased electrical pins on a connector between the launch vehicle and the ground equipment. Because the electrical circuits connected the spacecraft to the offending stage, this interface design flaw led to a spacecraft failure three days later.[15]

Some farsighted managers and engineers recognized that interfaces represented the connection not simply between hardware but also between individuals and organizations. Differences in organizational cultures, national characteristics, and social groups became critical when these groups had to work together to produce an integrated product. As the number of organizations grew, so too did the problems of communication. Project managers and engineers struggled to develop better communication methods.

As might be expected, international projects had the most difficult problems with interfaces. The most severe example was ELDO's Europa I and Europa II projects. With different countries developing each of three stages, a test vehicle, and the ground and telemetry equipment, ELDO had to deal with seven national governments, military and civilian organizations, and national jealousies on all sides. Within one year after its official inception, both ELDO and the national governments realized that something had to be done about the "interface problem." An Industrial Integrating Group formed for the purpose could not overcome the inherent communication problems, and every one of ELDO's flights that involved multiple stages failed. All but one failed because of interface difficulties.[16]

By the early 1960s, systems engineers developed interface control documents to record and define interfaces between components. On the manned space projects, special committees with members from each contributing organization worked out interfaces between the spacecraft, the rocket stages, the launch complex, and mission operations. After the fledgling European Space Research Organisation began to work with American engineers and managers from Goddard Space Flight Center, the first letter from the American project manager to his European counterpart was a request to immediately begin work on the interface between the European spacecraft and the American launch vehicle.[17]

Systems management became the standard for missile and space systems because it addressed many of the major technical issues of rockets and spacecraft. The complexity of these systems meant that coordination and communication required greater emphasis in missile and space systems than they did in many other contemporary technologies. Proper communication helped to create better designs. However, these still had to be translated into technical artifacts, inspected and documented through rigid quality inspections and testing during manufacturing. Finally, the integrated system had to be tested on the ground and, if possible, in flight as well. The high cost and "nonreturn" of each missile and spacecraft meant that virtually every possible means of ground verification paid off, helping to avoid costly and difficult-to-analyze flight failures. All in all, the extremes of the space environment, automation, and the volatility of rocket fuels led to new social methods that emphasized considerable up-front planning, documentation, inspections, and testing. To be implemented properly, these social solutions had to satisfy the needs of the social groups that would have to implement them.

Systems Management and Its Promoters

Four social groups developed and spread systems management: military officers, scientists, engineers, and managers. All the groups promoted aspects of systems management that were congenial to their objectives and fought those that were not. For example, the military's conception of "concurrency" ran counter in a number of ways to the managerial idea of "phased planning," while the scientific conception of "systems analysis" differed from the engineering notion of "systems engineering." Academic working groups promoted by scientists and engineers conflicted with hierarchical structures found in the military and industry, and the working groups' informal methods frustrated attempts at hierarchical control through formal processes. The winners of these bureaucratic fights imposed new structures and processes that promoted their conceptions and power within and across organizations.

In the early 1950s, the prestige of scientists and the exigencies of the Cold War gave scientists and military officers the advantage in bureaucratic competition. Military leaders successfully harnessed scientific expertise through their lavish support of scientists, including the development of new laboratories and research institutions. Scientists in turn provided the military with technical and political support to develop new weapons.[18] The alliance of these two groups led to the dominance of the policy of concurrency in the 1950s.[19]

To the air force, concurrency meant conducting research and development in parallel with the manufacturing, testing, and production of a weapon. More generally, it referred to any parallel process or approach. Concurrency met the needs of military officers because of their tendency to emphasize external threats, which in turn required them to respond to those threats. Put differently, for military officers to acquire significant power in a civilian society, the society must believe in a credible threat that must be countered by military force. If the threat is credible, then military leaders must quickly develop countermeasures. If they do not, outsiders could conclude that a threat does not exist and could reduce the military's resources. For the armed forces, external threats, rapid technological development, and their own power and resources went hand in hand.

Scientists also liked concurrency, because they specialized in the rapid creation of novel "wonder weapons" such as radar and nuclear weapons. Even when scientists had little to do with major technological advances, as in the case of jet and rocket propulsion, society often deemed the engineers "rocket scientists." Scientists did little to discourage this misconception. They gained prestige from technical expertise and acquired power when others deemed technical expertise critical. Scientists predicted and fostered novelty because discovery of new natural laws and behaviors was their business. Novelty required scientific expertise, whereas "mundane" developments could be left to engineers.

While the Cold War was tangibly hot in the late 1940s and 1950s, American leaders supported the search for wonder weapons to counter the Communist threat. Although very expensive, nuclear weapons were far less expensive than maintaining millions of troops in Europe, and they typified American preferences for technological solutions.[20] Military officers allied with scientists used this climate to rapidly drive technological development.

By 1959, however, Congress began to question the military's methods because these weapons cost far more than predicted and did not seem to work.[21] Embarrassing rocket explosions and air-defense system failures spurred critical scrutiny. Although Sputnik and the Cuban Missile Crisis dampened criticism somewhat, military officers had a difficult time explaining the apparent ineffectiveness of the new systems. Missiles that failed more than half the time were neither efficient military deterrents nor effective deterrents of congressional investigations. The military needed better cost control and technical reliability in its missile programs. Military officers and scientists were not particularly adept in these matters. However, managers and engineers were.

Engineers can be divided into two types: researchers and designers. Engineering researchers are similar to scientists, except that their quest involves technological novelty instead of "natural" novelty. They work in academia, government, and industrial laboratories and have norms involving the publication of papers, the development of new technologies and processes, and the diffusion of knowledge. By contrast, engineering designers spend most of their time designing, building, and testing artifacts. Depending upon the product, the success criteria involve cost, reliability, and performance. Design engineers have little time for publication and claim expertise through product success.

Even more than design engineers, managers pay explicit heed to cost considerations. They are experts in the effective use of human and material resources to accomplish organizational objectives. Managers measure their power from the size and funding of their organizations, so they have conflicting desires to use resources efficiently, which decreases organizational size, and to make their organizations grow so as to acquire more power. Ideally, managers efficiently achieve objectives, then gain more power by acquiring other organizations or tasks. Managers, like engineers, lose credibility if their end products fail.

As ballistic missiles and air-defense systems failed in the late 1950s, military officers and aerospace industry leaders had to heed congressional calls for greater reliability and more predictable cost. In consequence, managerial and engineering design considerations came to have relatively more weight in technology development than military and scientific considerations. Managers responded by applying extensive cost-accounting practices, while engineers performed more rigorous testing and analysis. The result was not a "low cost" design but a more reliable product whose cost was high but predictable. Engineers gained credibility through successful missile performance, and managers gained credibility through successful prediction of cost. Because of the high priority given to and the visibility of space programs, congressional leaders in the 1960s did not mind high costs, but they would not tolerate unpredictable costs or spectacular failures.

Systems management was the result of these conflicting interests and objectives. It was (and is) a mélange of techniques representing the interests of each contributing group. We can define systems management as *a set of organizational structures and processes to rapidly produce a novel but dependable technological artifact within a predictable budget.* In this definition, each group appears. Military officers demanded rapid progress. Scientists desired novelty. Engineers wanted a dependable product. Managers sought predictable costs. Only through successful collaboration could these goals be attained. To succeed in the Cold War missile and space race, systems management would also have to encompass techniques that could meet the extreme requirements of rocketry and space flight.

Conclusion

Social and technical concerns drove the development of systems management. The dangers of the Cold War fed American fears of Communist domination, leading to the American response to ensure technological superiority in the face of the quantitative superiority of Soviet and Chinese military forces. Military officers and scientists responded to the initial call by creating nuclear weapons and ballistic missiles as rapidly as possible.

Technical issues then reared their ugly heads, as the early missile systems exploded and failed frequently. Investigation of the technical issues led to the creation of stringent organizational methods such as system integration and testing, change control, quality inspections and documentation, and configuration management. Engineers led the development of these new technical coordination methods, while managers intervened to require cost and schedule information along with technical data with each engineering change.

The result of these changes was systems management, a mix of techniques that balanced the needs and issues of scientists, engineers, military officers, and industrial managers. While meeting these social needs, systems management also addressed the extreme environments, danger, and automation of missile and space flight technologies. By meeting these social and technical needs, systems management would become the standard for large-scale technical development in the aerospace industry and beyond.

TWO

Creating Concurrency

> We are in a technological race with the enemy. The time scale is incredibly compressed. The outcome may decide whether our form of government will survive. Therefore, it is important for us to explore whether it is possible to speed up our technology. Can we for example plan and actually *schedule* inventions? I believe this can be done in most instances, provided we are willing to pay the price and make no mistake about it, the price is high.
>
> —Colonel Norair M. Lulejian, 1962

The complex weapon systems of World War II and the Cold War involved enormous technical difficulties. Scale was not the problem, for large-scale systems such as the telephone network, electrical power systems, and skyscrapers had existed before. Rather, the difficulty lay in the heterogeneity of the components, their novelty, and their underlying complexity. Military personnel were unfamiliar with the new technologies of rocket engines, nuclear weapons, and guidance and control systems.

New technology provided opportunities for military officers with a technical bent. Allied with scientists and research engineers, these officers promoted the "air force of the future" over the traditional "air force of the present." Through wide-ranging research and fast-paced development, the air force would maintain a technological edge over its Communist adversaries. Separating research and development (R&D) from current operations, these officers created new methods to integrate technologies into novel "weapon systems." In so doing, they brought into being new organizations and niches for technical officers, scientists, and engineers.

Of the new technologies developed during World War II, ballistic missiles were among the most promising. The marriage of ballistic missiles with fusion warheads promised an invulnerable delivery system for the ultimate explosive. At the push of a button, an entire city could be obliterated within thirty minutes. While the bomber pilots who dominated the air force's leadership vacillated, technical officers and their scientific allies pressed ahead and past air force skeptics, winning top-priority status for intercontinental ballistic missiles (ICBMs). Led by Brig. Gen. Bernard Schriever, their success was the apex of scientific influence in the military and laid the foundation for a new way of organizing R&D. Combining scientific novelty with the military's need for rapid development, this new approach became known as concurrency.[1]

Concurrency replaced the air force's prior management methods for large-scale technology development. If the technology of ICBMs had been less complex, or if their development had occurred at a more relaxed pace, then the air force's existing management techniques might have sufficed. Facing the combined impact of technical difficulty and rapid tempo, however, the loosely organized technical divisions of the air force's development groups could not cope. Equally important, the scientists who advised the air force's leaders did not believe that traditional methods and organizations would succeed. Based on their recommendations, Schriever created a centralized, tightly planned management scheme to implement the air force's complex new weapon system as quickly as possible. To understand the changes that Schriever and his allies wrought, we must turn to the air force's methods prior to the development of ICBMs.

Aircraft before Systems

The air force's R&D methods trace back to the creation of aircraft in the first decade of the twentieth century. Because the army did not create an arsenal to develop aircraft, contractual relationships between the Army Air Corps and the aircraft industry governed military aircraft development. The Army Signal Corps ordered its first aircraft from the Wright brothers in 1908 using an incentive contract that awarded higher fees for a higher-speed aircraft.[2] Army evaluation and testing of aircraft began near the Wrights' plant in Dayton, Ohio. These facilities soon grew into the Air Corps's primary complex for aircraft development and testing.

While European powers rapidly developed aircraft for military purposes, the U.S. Army kept aircraft development a low priority. World War I broke American lethargy; in 1915, Congress created the National Advisory Committee for Aeronautics (NACA) to promote aircraft research, evaluation, and development for the military and the aircraft industry. Engineers at NACA's facility at Langley Field in Hampton, Virginia, concentrated on the testing and evaluation of aerodynamic structures and aircraft performance, using new wind-tunnel facilities to test fuselages, engine cowlings, propeller designs, and pilot-aircraft controllability. The United States mass-produced a few European designs during the war but rapidly dismantled most of its capability after the war's end.[3]

Between World War I and World War II, the Army Air Corps fostered aircraft development at a leisurely pace. Typically the engineering and procurement divisions at Wright Field in Dayton contracted with industry for aircraft, which officers, civilians, and operational commands then tested. Army Ordnance and the Army Signal Corps developed the armaments and electronic gear that Wright Field personnel then integrated into the aircraft. Wright Field procured the components, then modified them as necessary to integrate them into the aircraft. Funding constraints were more important than schedule considerations, leading to a rather deliberate development and testing program commonly described as the "fly before you buy" concept.[4]

After the Air Corps released design specifications, contractors designed, built, and delivered a prototype known as the X-model to the Air Corps. The Air Corps tested this model, making recommendations for changes. After completion of X-model testing, the contractor made the recommended design changes, then developed the Y-model production prototype. The Air Corps then ran another series of tests and made further design recommendations. After approval of the Y-model, the contractor released the production drawings and built the required number of aircraft.[5]

From the mid-1920s, Wright Field assigned a project engineer from its Engineering Division to monitor all aircraft design and development. By the late 1930s, Wright Field assigned a project officer to each aircraft in development, along with the project engineer and a small supporting staff. For example, in the Bombardment Branch before World War II, Col. Donald Putt and five other officers managed six aircraft projects with the assistance of a few secretaries and Wright Field engineers assigned to tasks as needed. Be-

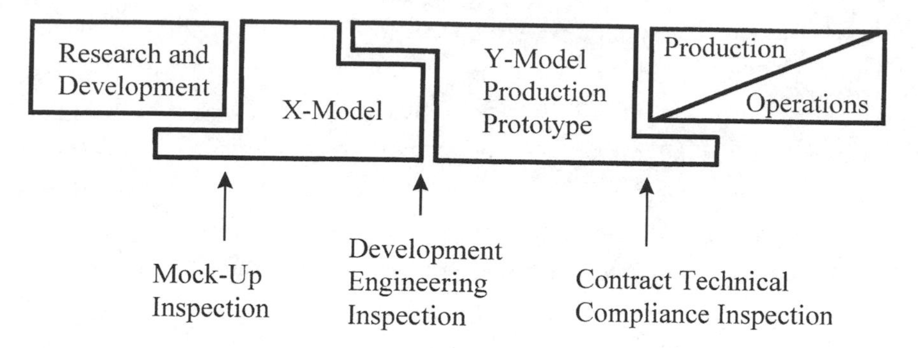

"Fly before you buy" sequential development, typical of the Army Air Corps in the 1920s and 1930s. Adapted from Benjamin N. Bellis, L/Col USAF Office DCS/Systems, "The Requirements for Configuration Management During Concurrency," in AFSC Management Conference, Air Force Systems Command, Andrews Air Force Base, Washington, D.C., AFHRA Microfilm 26254, 5-24-2.

cause of the slow pace of development, the limited role of the government in testing and approving designs, and the fixed-price contracting method typical before the war, this small staff sufficed. Project officers focused on aircraft safety and on finding design weaknesses.[6]

As war loomed in 1940, Congress legalized negotiated cost-plus-fixed-fee (CPFF) production contracts. With a flood of funding and a goal of building 50,000 aircraft, the Air Corps immediately signed letters of intent to get design and production moving, with cost negotiations deferred until later. Under the prior competitive bidding process, procurement officers did not need to understand the financial details of a manufacturers' bid, because the manufacturer—not the government—lost money if it underbid. However, under CPFF arrangements, cost overruns were the government's problem. The Air Corps Procurement Branch grew rapidly to collect information and negotiate with contractors to assess the validity of cost charges and determine a fair profit.[7]

Unless Congress extended the authority to negotiate contracts after the war, the military's capability to control industry and influence scientists and their new technologies would dramatically decrease. Fortunately for the military, the Procurement Act of 1947 extended the military's wartime authority and tools, including the formerly controversial negotiated contract mechanism, into peacetime.

The importance of the 1947 act should not be underestimated, for it perpetuated government use of CPFF contracts. This had several significant ramifications. First, the CPFF contract reduced risk for industry. Where high risk was inherent, as it was in R&D, this drew profit-making corporations and universities into government-run activities. Second, to reduce government risk, CPFF contracts required a government bureaucracy sufficient to monitor contractors. Third, CPFF contracts turned attention from cost concerns to technical issues. This "performance first" attitude led to higher costs but also to a faster pace of technical innovation and occasionally to radical technological change. Last, the CPFF contract provided some military officers with the means to promote technological innovation along with their own careers.[8]

Negotiated contracts formed the basis for Cold War contractual relationships between government, industry, and academia. Government officials became both partners and controllers of the aircraft industry in a way unimagined before the war, with expanded procurement organizations that made the federal government a formidable negotiator. To fully exploit their extended authority to create new weapons, however, the Army Air Forces would also have to solidify its relationships with scientists and engineers.

Organizing to Communicate with Technologists

During World War II, scientists vastly increased the fighting capability of both Allied and Axis powers. The atomic bomb, radar, jet fighters, ballistic missiles, and operations research methods applied to fighter and bomber tactics all had significant impact on the war. Recognizing the contributions of scientists, Gen. H. H. "Hap" Arnold, commander of the Army Air Forces, advocated maintaining the partnership between military officers and scientists after the war's end. His plans led to the creation of several organizations that cemented the partnership between technically minded Army Air Forces officers and the community of scientific and technological researchers.

In 1944, Arnold met briefly with eminent aerodynamicist Theodore von Kármán of the California Institute of Technology and asked him to assemble a group of scientists to evaluate German capabilities and study the Army Air Forces' postwar future. Among the group's recommendations were the establishment of a high-level staff position for R&D, a permanent board of scientists to advise the Army Air Forces, and better means to educate Army Air Forces officers in science and technology.[9] The Army Air Forces acted first to maintain the services of von Kármán and his scientific friends. Supported by General Arnold, the Army Air Forces established the Scientific Advisory Board (SAB) in June 1946 as a semipermanent adviser to the staff.[10]

Arnold recognized that establishing an external board of scientists would do little to change the Army Air Forces unless he also created internal positions to act as bridges and advocates for scientific ideas. He established the position of scientific liaison in the air staff and elevated his protégé Col. Bernard Schriever into the position in 1946. Schriever had known Arnold since 1933, when as a reserve officer Schriever was a bomber pilot and maintenance officer under Arnold. Schriever's mother became a close friend of Arnold's wife, leading to a lifelong friendship with the Arnold family. Arnold encouraged Schriever to take a full commission, which Schriever did prior to World War II. Schriever served with distinction in the Pacific, and his work in logistics brought him into contact with procurement officers at Wright Field. After the war, Arnold moved him to the Pentagon. As scientific liaison, Schriever helped create the air force's R&D infrastructure, including test facilities at Cape Canaveral, Florida, and in the Mojave Desert north of Los Angeles as well as research centers in Tennessee and near Boston. He worked closely with the SAB, an association that would have far-reaching consequences.[11]

Despite the creation of a research office in Air Materiel Command (AMC),[12] an increasing number of military officers believed that AMC did not pursue R&D with sufficient vigor. The controversy revolved around the conflict between technologically oriented officers who promoted the "air force of the future" and the traditional pilots who focused on the "air force of the present." Advocates of the future air force had powerful allies in General Arnold and in Lt. General Donald Putt, a longtime aircraft procurement officer from Wright Field. Putt had been a student of von Kármán at Caltech and in the late 1940s was director of R&D in the air force headquarters staff.[13]

Putt and an energetic group of colonels under him discussed how to improve air force R&D, which in their opinion languished in AMC. As budgets shrank after the war, AMC gave high priority to maintaining operational forces, leading to R&D budget cuts. This concerned members of the SAB as well as Putt's allies. Putt and his colonels plotted how the SAB could aid their cause.[14]

Capitalizing on an upcoming meeting of the SAB in the spring of 1949, Putt asked the chief of the Air Staff, Gen. Hoyt Vandenberg, to speak to the board. Vandenberg agreed, but only if Putt would write his speech. This was the opportunity that Putt and his protégés sought. Putt asked one of his allies, SAB military secretary Col. Ted Walkowicz, to write the speech. Walkowicz included "a request of the Board to study the Air Force organization to see what could be done to increase the effectiveness of Air Force Research and Development." Putt "rather doubted that Vandenberg would make that request." Fortunately for Putt, Vandenberg at the last minute backed out and had his deputy, Gen. Muir Fairchild, appear before the board. Fairchild, an advocate of R&D, read the speech all the way through, including the request. Putt had already warned SAB Chairman von Kármán what was coming, so von Kármán quickly accepted the request.[15]

Putt and his colonels knew that this was only the first step in the upcoming fight. They also had to ensure that the report would be read. Putt's group carefully picked the SAB committee to include members that had credibility in the air force. They selected as chairman Louis Ridenour, well known for his work on radar at the Massachusetts Institute of Technology's Radiation Laboratory. More important was the inclusion of James Doolittle, the famed air force bomber pilot and pioneer aviator who was also Vandenberg's close friend. Putt persuaded Doolittle to go on a duck hunting trip with Vandenberg after Ridenour and von Kármán presented the study results to the Air Staff. Putt later commented that "this worked perfectly," gaining the chief's ear and favor. Putt's group also coordinated a separate air force review to assess the results of the scientific committee. After hand-picking its members as well and ensuring coordination with Ridenour's group, Putt noted that "strangely enough, they both came out with the same recommendations."[16]

The Ridenour Report charted the air force's course over the next few years. It recommended the creation of a new command for R&D, a new graduate study program in the air force to educate officers in technical matters, and improved career paths for technical officers. The report also recommended the creation of a new general staff position for R&D separated from logistics and production, and a centralized accounting system to better track R&D expenditures. After a few months of internal debate, General Fairchild approved the creation of Air Research and Development Command (ARDC), separating the R&D functions from AMC. Along with ARDC, Fairchild approved creation of a new Air Staff position, the deputy chief of staff, development (DCS/D).[17]

With the official establishment of ARDC and the DCS/D on January 23, 1950, the air force completed the development of its first organizations to cement ties between technically minded military officers and scientific and technological researchers. These new organizations, which also included the RAND Corporation,[18] the Research and Development Board (RDB),[19] and the SAB, would in theory make the fruits of scientific and technological research available to the air force. The RDB and SAB coordinated air force efforts with the help of the scientists and engineers, similar to how the wartime Office of Scientific Research and Development had operated, but RAND was a new kind of organization, a "think tank." ARDC and the DCS/D would attempt to centralize and control the air force's R&D efforts. They would soon find that for large projects, they would have to centralize authority around the project, instead of the technical groups of AMC or ARDC.

The Rise of the Weapon System Concept

The air force had to develop two kinds of technologies. The majority of the projects were concerned with component development. On account of their great cost and complexity, however, large-scale weapons such as bombers, fighters, and missiles took up the bulk of the air force's R&D resources. To manage these so-called weapon systems, air force officers found that their loosely organized prewar methods did not suffice. For the new systems, the air force looked to new models of centralized project management.

Two World War II aircraft projects fit the bill. The complex B-29 and P-61 projects both used committees to coordinate the development of the airframe, electronics, and armament during development, instead of after airframe manufacture and testing.[20] For the complex and pressurized *B-29*, engineers designed armament and communications together from the start, because the aircraft's computer-controlled fire-control systems were integrally connected to the airframe. For the *B-29* and the *P-61*, officers considered the entire aircraft a system that included manufacturing and training as well as hardware.[21]

Another influential World War II program was the Manhattan Project to build the atomic bomb. Gen. Leslie Groves of the Army Corps of Engineers managed the project, gathering physicists, chemists, and engineers at Los Alamos, New Mexico, to design the bomb. Groves administered the project with a staff of three and made major decisions with a small committee consisting of himself, Vannevar Bush, James Conant, and representatives of each of the services. Army officers directed day-to-day operations at each of the project's field sites, most of which had traditional hierarchical organizations, albeit cloaked in secrecy. Because of technical and scientific uncertainties, the project developed two bomb designs and three methods to create the fissile material.[22]

The organization at Los Alamos differed from the organization at other project sites. Director Robert Oppenheimer wrested a degree of freedom of speech for the scientists and ensured that they remained civilians. Oppenheimer, to respect the traditional independence of scientists and maintain open communication, initially adopted the loose department structure of universities. This changed in the spring of 1944, when tests showed that the plutonium gun assembly bomb would not work. The tests led to an acceleration of work on the more complex implosion design. As R&D teams grew, the project needed and obtained strong managers like Robert Bacher and George Kistiakowsky, who transformed the project's organization from an academic model to divisions organized around the end-product—a project organization.[23]

Americans also learned from the organization of the German V-2 project, headed by Wernher von Braun. Reporting to General Arnold on German scientific capabilities at the end of World War II, von Kármán stated that one of the major factors in the success of the German V-2 project was its organization:

> Leadership in the development of these new weapons of the future can be assured only by uniting experts in aerodynamics, structural design, electronics, servomechanisms, gyros, control devices, propulsion, and warhead under one leadership, and providing them with facilities for laboratory and model shop production in their specialties and with facilities for field tests. Such a center must be adequately supported by the highest ranking military and civilian leadership and must be adequately financed, including the support of related

> work on special aspects of various problems at other laboratories and the support of special industrial developments. It seems to us that this is the lesson to be learned from the activities of the German Peenemünde group.[24]

In the Ridenour Report of 1949, the SAB remembered the lessons of the Manhattan and V-2 projects for organizing large new technologies. They noted that new systems were far more complex than their prewar counterparts, making it necessary for some engineers to concentrate on the entire system instead of its components only. Project officers also needed greater authority to better lead a task force of "systems and components specialists organized on a semi-permanent basis." Because the air force had few qualified technical officers, the committee recommended that the air force draw upon the "very important reservoir of talent available for systems planning in the engineering design staffs of the industries of the country."[25]

Despite the recommendations of the Ridenour Report, AMC officers at Wright Field continued to organize projects on functional lines mirroring academic disciplines and to coordinate projects through small project offices. As late as 1950, typical project offices had fewer than ten members, and engineering expertise, parceled out from Wright Field's functional divisions, were, as one historian put it, "only casually responsible" to the project office.[26] At the time, AMC's Col. Marvin Demler stated: "Due to the complexity of the mechanisms which we develop, and our organization by hardware specialties, a very high degree of cooperation and coordination is required between organizations at all levels. In fact, an experienced officer or civilian engineer coming to Wright Field for the first time simply cannot be effective for perhaps six months to one year while he learns 'the ropes' of coordination with other offices. The communication between individuals necessary for the solution of our problems of coordination defy formal organizational lines."[27]

For large projects, this informal structure was not to continue for much longer. When the Korean War broke out in late 1950, the air force found itself with numerous unusable aircraft. In January 1951, Vice Chief of Staff Nathan Twining instructed DCS/D Gen. Gordon Saville to investigate the air force's organization to determine whether it contributed to the poor aircraft readiness. Saville ordered the formation of a study group, led by Colonel Schriever, to investigate the problem. The group returned to the comments of the Ridenour Report regarding the lack of technical capability in the air force and the problems caused by separating airframe development from component development.[28]

Schriever's group completed its study in April 1951 and released an influential paper called "Combat Ready Aircraft." It pinpointed two major problems with current aircraft: requirements based on short-term factors, leading to continuous modifications, and insufficient coordination and direction of all elements of the "complete weapon."[29] The latter concern probably arose from the Ridenour Report and the examples of the *B-29*, the *V-2*, and the Manhattan Project.

To solve these problems, the group recommended that the air force create an organization and process with responsibility and authority over the complete weapon by adding "planning, budgeting, programming, and control" to the functions of the responsible air force organizations. The organizations would have complete control over the entire projects, enforced through full budget authority.[30] Examples of this kind of organization already existed in the air force's guided missile programs. These weapons differed substantially from piloted aircraft, and the separate procurement of airframe, engines, and armament (payload) made little sense.[31] The study group suggested that the air force let prime contracts to a single contractor to integrate the entire weapon and that the air force organize on a project basis.

Changes to the procurement cycle had to be addressed as well. The group noted that in World War II, decisions to produce aircraft occurred haphazardly and that aircraft rolled off the assembly line directly to combat units at the same time as they were delivered to testing. Because production continued rapidly and little testing occurred, invariably the operational units found numerous problems, leading to the grounding of aircraft for modifications. Believing the current emergency did not allow for the fly-before-you-buy sequential approach and that the delivery of the production aircraft to combat units was dangerous and wasteful, the group selected a solution that was a compromise. It recommended eliminating the X- and Y-model aircraft but slowing the initial production line until test organizations found and eliminated design bugs. Only then should production be accelerated, it said. The air force would select contractors based on the best proposal instead of through a "fly-off" of aircraft prototypes. These ideas, along with project-centered organization and simultaneous planning of all components throughout the weapon's life cycle, defined the weapon system concept.[32]

Brigadier General Putt, now commander of Wright Air Development Center, immediately campaigned for the weapon system concept among the component developers at Wright Field. He had a difficult sell because the new organization had moved power from the functional organizations to the project offices. The project office was to act on a systems basis, making compromises between cost, performance, quality, and quantity. Putt admonished the component engineers: "Somebody has to be captain of the team, and decide what has to be compromised and why. And that responsibility we have placed on the project offices." He also stated in no uncertain terms who had the authority, telling the component engineers that they needed to be "sure that all the facts" had "been placed before" the project office. "At that time," he told the engineers, "your responsibility ceases."[33]

Without a large number of technical officers, the air force handed substantial authority to industry. Under the weapon system concept, the air force "purchased *management* of *new* weapon system development and production." However, contractors had to "accept the Air Force as the monitor of his [the contractor's] plans and progress, with the cautionary power of a partner and the final veto power of the customer." The air force stated that it could not "escape its own responsibility for system management simply by assigning larger blocks of design and engineering responsibility to industry." Although the new process gave industry a larger role, air force officers would not remain passive.[34]

Adoption of the weapon system concept throughout the air force did not go smoothly, because of continuing disagreements between the DCS/D and ARDC on one hand, and the deputy chief of staff, materiel (DCS/M), and AMC on the other. The key question that divided the fledgling ARDC and its parent, AMC, was when "development" ended and "production" began. If production started early in a weapon's life cycle, then AMC maintained greater control, whereas if development ended relatively late in the cycle, then ARDC acquired more power. Not surprisingly, AMC leaned toward a definition of production that encompassed earlier phases of the life cycle, while ARDC opted for late-ending development. Because development continued as long as changes to the weapon occurred, and because production began

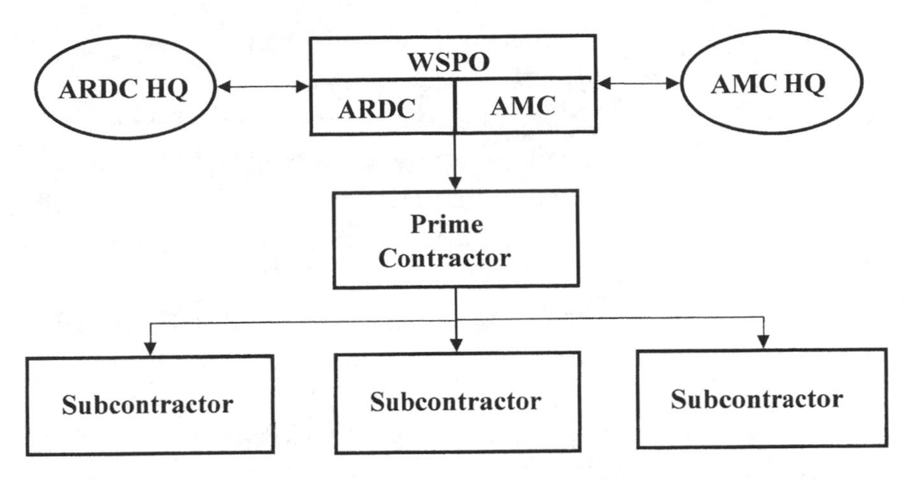

Weapon System Project Office implementation of the "system concept."

the moment the first prototype was built, no objective definition tipped the scales one way or another. Under such circumstances, the air force's official arbitrator between ARDC and AMC, James Doolittle, had to intervene.

In April 1951, Doolittle reported that because development continued through a system's entire life cycle, the ARDC definition should hold. In consequence, ARDC should control production engineering.[35] The new agreement led to the issuance of Air Force Regulation (AFR) 20-10, "Weapon System Project Offices," in October 1951. The regulation specified that every major project should have a Weapon System Project Office (WSPO), with officers from ARDC and AMC in charge.

A marvel of diplomacy, the document stated that during the early portions of development, the ARDC representative would be the "team captain," and in the later portions, after a decision to produce the article in quantity, the AMC representative would be the "team captain." In practice, the line between the two was fuzzy, leaving the two officers to work it out for themselves based on circumstances or personalities. The team captain coordinated the activities for the entire project but did not have authority over the other officer. If the two could not agree, they would both have to take the problem to higher authorities, potentially all the way up to the DCS/D and DCS/M at air force headquarters.[36]

The resulting ambiguities continued to cause organizational headaches, leading once again to intervention by Doolittle. This time Doolittle did not feel comfortable forcing a solution, so he recommended another Air Staff study to investigate the problem. His only proviso was that the group protect the importance of R&D. The Air Staff gave the DCS/M, Lt. General Orval Cook, responsibility for solving the interface problems. In cooperation with DCS/D Laurence Craigie, Cook appointed a task group, the "Cook-Craigie Group," to work on the issue. Group members decided that ARDC should keep responsibility for weapon systems until the Air Staff stated in writing that the weapon should be purchased.[37] The new process, known as the Cook-Craigie Procedures of March 1954 and formalized by modification of AFR 20-10 in August of that year, momentarily ended the bickering between the development and materiel groups. Their unity would be tested severely with the development of the air force's most radical new weapon, ballistic missiles.

ICBMs and Formation of the Inglewood Complex

Missiles, particularly ballistic missiles,[38] disrupted the air force's culture, operations, and organization in several important ways. First, and most obviously, missiles had no pilots, relegating humans to only "pushing a button." Second, maintenance and long-term operations of missiles amounted to storage and occasional refurbishment, as opposed to the ongoing repairs typical for aircraft. Third, because missiles were used just once, missile testing required the creation of a missile production line. Unlike aircraft, where a few prototypes could be built and tested with dozens or hundreds of flights each, every missile test required a new missile. This implied that the fly-before-you-buy concept, where aircraft could be tested before instigation of full-scale production, no longer applied. For missiles, testing required a production line. Finally, missiles involved a variety of challenging new technical issues, as described in the previous chapter. Simply put, many of the air force's existing organizational and technical processes did not work for missiles.

Ballistic missile programs languished at a low priority during and after World War II, as the air force concentrated its efforts first on manned bombers, and then on jet fighters for the Korean War.[39] The rapidly escalating Cold War provided the impetus to transform the loosely organized missile projects. Successful testing of the Soviet atomic bomb in 1949 spurred the United States to develop a fusion weapon. In March 1953, Assistant for Development Planning Bernard Schriever learned of the success of American thermonuclear tests from the SAB. Recognizing the implications of this news, within days Schriever met renowned mathematician John von Neumann at his Princeton office. Von Neumann predicted that scientists would soon develop nuclear warheads of small enough size and large enough explosive power to be placed on ICBMs. Because of their speed and in-flight invulnerability, ICBMs were the preferred method for nuclear weapons delivery, if the air force could make them work. Realizing that he needed official backing, Schriever talked with James Doolittle, who approached Chief of Staff Vandenberg to have the SAB investigate the question.[40]

The Nuclear Weapons Panel of the SAB, headed by von Neumann, reported to the air force staff in October 1953. In the meantime, Trevor Gardner, assistant to the secretary of the air force, volunteered to head a Department of Defense (DOD) Study Group on Guided Missiles. Gardner learned of Convair's progress on its *Atlas* ICBM and met with Dr. Simon Ramo, an old friend and head of Hughes Aircraft Company's successful air-to-air missile project, the Falcon. Based on the results of his study group, Gardner and Air Force Secretary Talbott formed the Strategic Missiles Evaluation Committee, or Teapot Committee, to recommend a course of action for strategic ballistic missiles.[41]

Von Neumann headed the group, and Gardner selected Ramo's newly created Ramo-Wooldridge Corporation (R-W) to do the paperwork and manage the day-to-day operations of the study. Ramo had partnered with fellow Hughes manager Dean Wooldridge to form R-W.[42] In February 1954 the Teapot Committee recommended that ICBMs be developed "to the maximum extent that technology would allow." It also recommended the creation of an organization that hearkened back to the Manhattan Project and Radiation Laboratory of World War II: "The nature of the task for this new agency requires that over-all technical direction be in the hands of an unusually competent group of scientists and engineers capable of making systems analyses, supervising the research phases, and completely controlling the experimental and hardware phases of the program—the present ones as well as the subsequent ones that will have to be initiated."[43]

On May 14, 1954, the air force made Convair's *Atlas* its highest R&D priority. Because Convair and the majority of the aircraft industry hailed from

Southern California, the air force established its new ICBM development organization, the Western Development Division (WDD), in a vacant church building in Inglewood, near Los Angeles airport. Air force leaders placed newly promoted Brig. General Bernard Schriever in command on August 2, 1954. Because the Teapot Committee had recommended creation of a "Manhattan-like" project organization, one of Schriever's first tasks was to see if this made sense and determine who would oversee the technical aspects of the project.[44]

Schriever rejected the Manhattan Project organization because ICBMs were significantly more complicated than the atomic bomb.[45] Because neither he nor the scientists believed that the air force had the technical expertise to manage the program, Schriever could hire Convair as prime contractor, or he could hire R-W as the system integrator, with Convair and other contractors as associate contractors. The air force used the prime contractor procedure on most programs, but this assumed that the prime contractor had the wherewithal to design and build the product. Schriever was already unhappy with Convair because he believed Convair kept "in-house" elements such as guidance and electronics in which it had little experience, to the program's detriment.[46]

Scientists with whom he had worked for nearly a decade also deeply influenced Schriever. Von Neumann and his fellow scientists believed the Soviet threat required a response like the Manhattan Project a decade earlier, bringing together the nation's best scientists to marry ballistic missiles to thermonuclear warheads. Schriever later explained:

> Complex requirements of the ICBM and the predominant role of systems engineering in insuring that the requirements were met, demanded an across-the-board competence in the physical sciences not to be found in existing organizations. Scientists rated the aircraft industry relatively weak in this phase of engineering, which was closely tied to recent advances in physics. The aircraft industry, moreover, was heavily committed on major projects, as shown by existing backlogs. Its ability to hire the necessary scientific and engineering talent at existing pay-scales was doubted, and with the profit motive dominant, scientists would not be particularly attracted to the low-level positions accorded to such personnel in industry.[47]

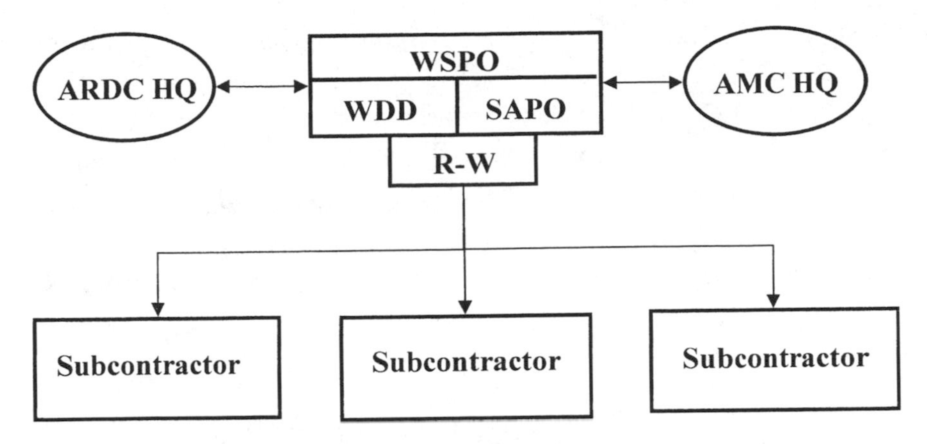

Organization of the Inglewood complex: the Western Development Division, the Special Aircraft Projects Office, and Ramo-Wooldridge.

Many years later Schriever described his admiration of the scientists: "I became really a disciple of the scientists who were working with us in the Pentagon, the RAND Corporation also, so that I felt very strongly that the scientists had a broader view and had more capabilities. We needed engineers, that's for sure, but engineers were trained more in a, let's say a narrow track having to do with materials than with vision."[48]

To capitalize on the vision and expertise of physical scientists and mathematicians such as von Neumann and von Kármán, Schriever created an organizational scheme whereby the leading scientists could guide the ICBM program. Following the von Neumann committee recommendations, Schriever selected R-W for systems engineering and integration.[49] Free of civil service regulations, R-W could hire the requisite scientific and technical talent. The air force could more easily direct R-W than Convair, because R-W had few contracts and no production capability. The aircraft industry disputed this unusual arrangement, fearing that it established a precedent for "strong system management control" by the air force and also that it might create a powerful new competitor with inside information about air force contracts and contractor capabilities. On both counts, the aircraft industry was correct.[50]

Selecting the best and brightest technical officers from ARDC, Schriever's talented staff quickly took charge of ICBM development. Because AMC retained procurement authority, it set up a field office known as the Special Aircraft Projects Office (SAPO) alongside Schriever's ARDC staff in Inglewood. By September 1954, air force headquarters approved Schriever's selection of R-W, confirming the triumvirate of the WDD, the SAPO, and R-W. Schriever's next battle would be to establish the authority and credibility of his team in the face of skepticism at air force headquarters and the outright hostility of the aircraft industry.[51]

Establishing the WDD's Authority

With Schriever's organizational foundations set, the immediate task was to push ICBM development rapidly forward and create a detailed plan within a year. Headquarters control and oversight would come through the budget process, so Schriever knew that until he had his plans worked out, he had to keep the budget profile low. He reallocated budgets from several air force organizations and was careful not to ask for too much at the start. Over the long haul, Schriever knew that the massive budget that he needed would require congressional appropriations and that he would have to vigorously defend his plan and its costs. To put off this day of reckoning, in October 1954 he requested a relatively small budget, realizing that there would have to be a major readjustment in the spring. "This support can be obtained by carefully planned and formalized action at the highest levels in the administration," he recognized. In this breathing space, he developed his technical plans, costs, justifications, and political strategy.[52]

Selection of Atlas contractors was the next task of Schriever's team. With the design still in flux, this would have to be done based on company capabilities instead of design competitions. Bypassing standard procurement regulations, Schriever ordered R-W to let subcontracts to potential suppliers to involve them in and educate them on the program. This allowed R-W to assess contractors as well as speed development and procurement. Schriever could not ignore all of the air force's procurement procedures. He had his team create performance specifications and perform "prebidding activities" to prepare for a competitive bidder's conference. Because of the in-depth knowledge R-W had gained through its subcontracts, Schriever had R-W contribute to the Source Selection Boards, providing inputs as requested by the air force. This was a serious (and possibly illegal) departure from standard procurement policy, which required that only government officials control contractor selection.[53]

Schriever directed R-W and his air force team to reassess the *Atlas* design and to determine Convair's role. Convair, which had been developing *Atlas* since January 1946, understandably believed that it deserved the prime contract to build, integrate, and test the vehicle. It vigorously campaigned against Schriever and the upstart R-W. Convair's leaders sparred with Schriever's organization for the next few months before they resigned themselves to R-W's presence. To appease the air force's scientific advisers, and to gain electronics capability, Convair executives hired highly educated scientists and engineers. For his part, Schriever placed restrictions on R-W to maintain some semblance of support from the aircraft industry. In a memo dated February 24, 1955, the air force prohibited R-W from engaging in hardware production on any ICBM program in which it acted as the air force's adviser and systems engineer.[54]

R-W had three tasks: to establish and operate the facilities for the Inglewood complex, to assess contractor capabilities, and to investigate the *Atlas* design. R-W made its first important contribution in the design task. The required mass and performance of the missile depended upon the size of the warhead and the reentry vehicle, for small changes in their mass led to large changes in the required launch vehicle mass. Working with the Atomic Energy Commission and other scientists, R-W scientists and engineers found that a new blunt cone design decreased the nose cone's weight by half, from about 7,000 to 3,500 pounds. This in turn decreased required launch vehicle weight from 460,000 to 240,000 pounds and reduced the number of engines from five to three. This dramatic improvement discredited Convair's claim to expertise and convinced Schriever, his team, and his superior officers that the selection of R-W had been correct.[55]

The most significant technical issue facing Schriever's group in the fall and winter of 1954 was the uncertainty of the design. Group members simply could not predict which parts of the design would work and which might not. R-W had been investigating a two-stage vehicle, and the initial results looked promising. In March 1955, Schriever convinced Lt. General Thomas Power,

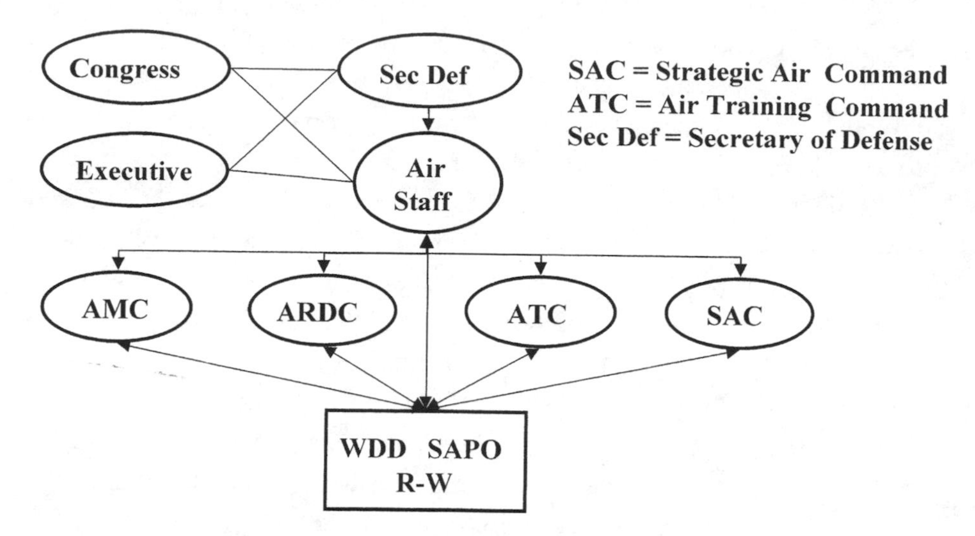

Pre-Gillette organization of ballistic missile development.

the ARDC commander, that a two-stage vehicle should be developed as a backup to *Atlas.* By May 1955, the WDD was working on *Atlas,* the two-stage *Titan,* and a tactical ballistic missile (ultimately known as *Thor*) as well.[56]

In the meantime, Schriever considered how best to fund the program. One possibility was to allocate the funds to a number of different budgets, then pull them back together in Schriever's group. This approach would hide the true budget amounts from effective oversight. However, the budgets required were too large to hide in this manner. With programmatic invisibility unlikely, Schriever's deputy, William Sheppard, argued that the best approach was to have a "separately justified and separately managed lump sum."[57]

Schriever had already discussed this approach with Gardner, and the two of them plotted a political strategy. Many of Schriever's budget actions required coordination with and justification to various organizations. Frustrated with the delays inherent in this coordination, Gardner and Schriever decided that they had to increase Schriever's authority and funding and decrease the number of organizations that could oversee and delay ICBM development. Both Schriever and Gardner recognized that they needed political support, so they vigorously sought it in Congress and within the Eisenhower administration. Gardner and Schriever briefed President Dwight D. Eisenhower in July 1955, eventually convincing him and Vice President Richard M. Nixon—with John von Neumann's timely support—to make ICBMs the nation's top defense priority.[58]

With the president's endorsement in hand by September, Schriever presented to Gardner the entire air force approval process, which required 38 air force and DOD approvals or concurrences for the development of ICBM testing facilities. Appalled, Gardner had him show it to Secretary of the Air Force Donald Quarles, who asked them to recommend changes to reduce the paperwork and delays. Gardner and Schriever formed a study group, loading it, as Schriever put it later, "pretty much with people who knew and who would come up with the right answers." Hyde Gillette, the deputy for budget and program management in the Office of the Secretary of Defense, chaired the group, which was to recommend management changes to speed ballistic missile development.[59]

Despite objections from AMC, which did not want to lose any more authority, the Gillette Committee agreed with Schriever that the multiple approvals and reporting lines caused months of delay. In consequence, the "Gillette Procedures," approved by Secretary of Defense Charles Wilson on

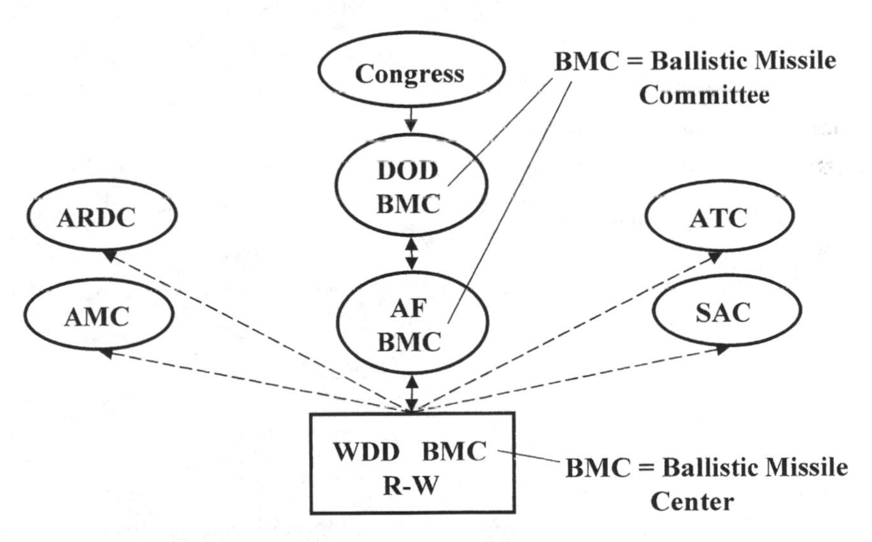

Ballistic missile organization—Gillette Procedures. *Solid lines with arrows* show the direct chain of authority. The air force's commands have no authority over ballistic missile development, and the Air Staff has input only through the Department of Defense Ballistic Missile Committee.

November 8, 1955, funneled all ballistic missile decisions through a single Ballistic Missile Committee in the Office of the Secretary of Defense. Although evading ARDC and AMC for approvals and decisions, Schriever's organization needed to provide them information. Schriever stated: "We had to give them information because they provide a lot of support, you see, so it wasn't the fact that we were trying to bypass them. We just didn't want to have a lot of peons at the various staff levels so they could get their fingers on it."[60] The Ballistic Missile Committee reviewed an annual ICBM development plan, and the Office of the Secretary of Defense would present, approve, and fund the ICBM program separately from the air force's regular procedures. In the development plan would be information on programming (linking plans to budgets), facilities, testing, personnel, aircraft allocation, financial plans, and current status. By 1958, AMC managers had trimmed industrial facility lead time from 251 to 43 days, showing the effectiveness of the new process.[61]

The Gillette Procedures relegated AMC, ARDC, and the operational commands to aiding the ICBM program, without the authority to change or delay it. From a parochial air force viewpoint, the only good thing about the program was that the completed missiles would eventually become part of the Strategic Air Command. Many in the air force did not take ballistic missiles seriously enough to fight for control over them. Col. Ray Soper, one of Schriever's trusted subordinates, noted that "the Ops [operational commands] attitude, at the Pentagon, was to let the 'longhairs' develop the system—they really didn't take a very serious view of the ballistic missile, for it was thought to be more a psychological weapon than anything else."[62]

With the adoption of the Gillette Procedures, Schriever garnered authority directly from the president, with a single approval of a single document each year required for ICBM development. Schriever's organization drew upon the best personnel and air force services, without having them interfere with his authority or decision processes. These new procedures represented the first full application of project management in the air force, where the project manager had both technical and budget authority for the project. Prior to this time, each project drew funds from several budgets and thus required separate justifications for each. The Gillette Procedures made the air force's financial and accounting system consistent with the authority of the project manager, although Gardner was unable to separate the ICBM budgets from the rest of the air force.[63] With these procedures in hand, Convair and the contractors under control, and the air force's regular bureaucracy shunted out of the way, Schriever drove the ICBM program at full speed, with little heed to cost, using the strategy of concurrency.

Concurrency

Rapid development of ICBMs required parallel development of all system elements, regardless of their technological maturity. Schriever called this *concurrency,* a handy word that meant that managers telescoped several typically serial activities into parallel ones. In serial development, research led to initial design, which led to prototype creation, testing, and manufacturing. Once the new weapon was manufactured, the operational units developed maintenance and training methods to use it. Under concurrency, these elements overlapped. Schriever did not invent the process but rather coined the term as a way of explaining the process to outsiders.[64]

Schriever's version of concurrency combined concepts learned over the previous decade. Parallel development had been practiced during World War II on the Manhattan and B-29 projects. Management structured around the product instead of by discipline had also been used on these projects. The combination of ARDC and AMC officers into a project-based office was a method applied since 1952, and Schriever's use of R-W to perform systems analyses like the *Atlas*'s nose cone design had also been foreshadowed by the RAND Corporation's development of systems analysis after World War II. Schriever claimed that concurrency was a new process. But was it?

One difference was that in the 1950s parallel development, once a wartime expedient, became a peacetime activity. With Congress exercising detailed oversight typical of peacetime, Schriever had to explain his processes in more detail than his wartime predecessors had. As Secretary of the Air Force James Douglas later told Congress, "I am entirely ready to express the view that . . . you have to subordinate the expenditure . . . to the urgency of looking to the end result." Or as Gardner succinctly stated, "We have to buy time with money." The term "concurrency" helped explain and justify their actions to higher authorities.[65]

A second major difference was in the nature of the technologies to be inte-

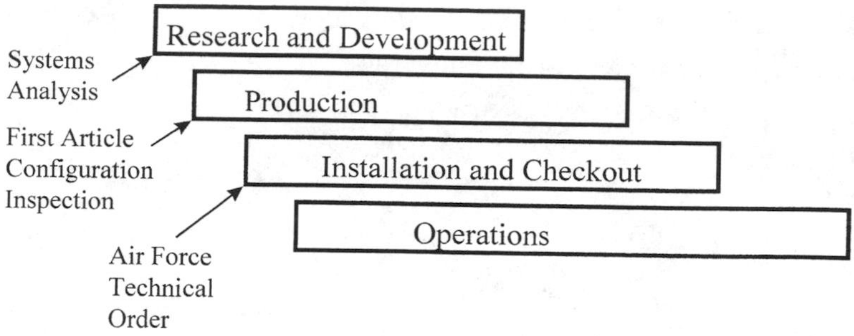

Research and Development
Production
Installation and Checkout
Operations
Systems Analysis
First Article Configuration Inspection
Air Force Technical Order

Concurrency. Adapted from Benjamin N. Bellis, L/Col USAF Office DCS/Systems, "The Requirements for Configuration Management During Concurrency," in AFSC Management Conference, Air Force Systems Command, Andrews Air Force Base, Washington, D.C., AFHRA Microfilm 26254, 5-24-3.

grated into ICBMs. In pre–World War II bombers, for example, engineers simply mounted machine guns at open side windows. However, with the *B-29* bomber, and for postwar aircraft, operators maneuvered machine guns with servomechanisms within a pressurized bubble, itself part of the airframe. Similarly, missiles had to be built with all elements planned and coordinated with each other from the start. Postwar weapons were far more complex than their prewar counterparts and more complex than the nuclear weapons of the Manhattan Project. Concurrency in the Cold War required far more detailed planning than previous concurrent approaches.

One application of concurrency was in selection of contractors for Atlas, and then for Titan and Thor. R-W performed the technical evaluations and gave input to ad hoc teams of WDD and SAPO personnel. The AMC-ARDC committees selected which companies they would ask to bid, evaluated the bids, and selected a second contractor for some subsystems. Selecting a concurrent contractor increased chances of technical success, stimulated better contractor performance by threatening a competitive contract if the first contractor performed poorly, and kept contractors working while the air force made decisions. To speed development, the SAPO issued letter contracts, deferring contract negotiations until later. In January 1955, the SAPO formalized the ad hoc committees, which became the AMC-ARDC Source Selection Board.[66]

To maximize flexibility and speed, Schriever initially organized the WDD with disciplinary divisions modeled on academia. Only in 1956 did the proliferation of projects lead him to create WSPOs for each project, consisting of AMC and ARDC representatives, as required by the weapon system concept. Until that time, most work occurred through ad hoc teams led by officers to whom Schriever had assigned the responsibility and authority for the task at hand. For example, when the WDD began to develop design criteria for facilities in March 1955, Schriever named Col. Charles Terhune, his technical deputy, "team captain" for the task. He also requested that R-W personnel assist. Terhune then led an ad hoc group to accomplish the task, and that group dissolved upon task completion.[67]

The fluid nature of the ad hoc groups and committees may well have maximized speed, but they also played havoc with standard procedures of the rest of the air force, which after all had to support ICBM development. Schriever initiated a series of coordination meetings with AMC, Strategic Air Command, air force headquarters, and other commands in December 1954. After the December meeting, the AMC Council decided it needed quarterly reports from the WDD to keep abreast of events. Over the next six months, AMC planning groups bickered with WDD personnel over reporting and support, as AMC needed information for personnel and logistics planning. AMC tried to plan tasks from Wright Field, whereas the WDD (and soon the SAPO) accomplished planning rapidly on-site, with little documentation or formality. AMC accused the WDD of refusing to provide the necessary data, whereas the WDD accused AMC officers of a lack of interest.

Disturbed because Schriever's crew had neither WSPOs nor Weapon System Phasing Groups (normally used to coordinate logistics), AMC had some reason to complain. As stated by the assistant for development programming, Brig. Gen. Ben Funk, "The normal organizational mechanisms and procedures for collecting and disseminating weapon system planning during the weapon system development phase did not exist," leading to gaps in the flow of information necessary for coordination. By the summer of 1955, SAPO personnel at the WDD made concerted efforts to pass information to AMC headquarters and to bring AMC planning information into the WDD.[68]

Schriever's need for speed led to extensive use of letter contracts through 1954 and 1955. Procurement officials in the SAPO and technical officers in the WDD realized that they needed to track expenditures relative to technical progress, but the rapid pace of the program and the lack of documentation quickly led to a financial and contractual morass. Complicated by the WDD's lack of personnel and the new process of working with R-W to issue technical directives, contractual problems became a major headache for the SAPO and AMC and another source of friction between Schriever and AMC leaders.[69]

The SAPO had authority to negotiate and administer contracts but initially lacked the personnel to administer them over the long term. Instead, SAPO personnel reassigned administration to the field offices of other commands "through special written agreements."[70] This complicated arrangement led to trouble. Part of the problem was the difficulty of integrating R-W into the management of the program. R-W had authority to issue contractually binding "technical directives" to the contractors, but instead of using these, R-W personnel sometimes "used the technical directive as a last resort, preferring persuasion first through either periodic meetings with contractor personnel or person-to-person visits between R-W and contractor personnel." This meant that many design changes occurred with no legal or contractual documentation. Because officers in the SAPO did not have enough personnel to monitor all meetings between R-W and the contractors and were not initially included in the "technical directive coordination cycle," matters soon got out of hand.[71]

This problem emerged during contract negotiations, as SAPO procurement officers and the contractors unearthed numerous mismatches between the official record of technical directives and the actual contractor tasks and designs. As differences emerged, costs spiraled upward, leaving huge cost overruns that could not be covered by any existing or planned funding. A committee appointed to investigate the problem concluded in June 1956 that "almost everyone concerned had been more interested in getting his work done fast than in observing regulations." It took the committee somewhat more than six months to establish revised procedures acceptable to all parties.[72]

The initial application of concurrency in Schriever's triad of the WDD, the SAPO, and R-W sped ICBM development but also spread confusion, disrupted communications with other organizations, and created a mountain of contractual, financial, and, as we shall see, technical problems. Flexible committees flicked in and out of existence, while supporting organizations outside of Schriever's group struggled to acquire the information they needed to assist. The strategy of parallel development, separated from the air force's normal routine, produced quick results, but the mounting confusion begged for a stronger management scheme than ad hoc committees.

Conclusion

World War II and the Cold War enabled the military to consolidate and extend its relationships with both academia and industry. When in 1947 the Procurement Act gave the DOD the permanent authority to negotiate contracts, military officers enlisted the support of academia and industry. Air force officers such as Hap Arnold, Donald Putt, and Bernard Schriever used scientists to create a technologically competent and powerful air force. Two models for relationships between the air force and the scientists evolved. First, RAND, the SAB, and the RDB continued the voluntary association of scientists with the military, as had occurred in World War II. However, the DCS/D and ARDC represented new air force efforts to gain control over the scientists through a standard air force hierarchy. Both models would continue into the future. Through these organizations and their personnel, air force officers hoped to develop the air force of the future.

When ICBMs became a possibility in late 1953, Schriever capitalized on his scientific connections, urging John von Neumann to head the Teapot Committee, which recommended that ICBMs be developed with the utmost speed and urgency. While Schriever and Assistant Secretary of the Air Force Trevor Gardner maneuvered behind the scenes to promote ICBMs, the Teapot Committee recommended the creation of a scientific organization on the Los Alamos model to recruit scientists to run the ICBM program. Unsure of the industry's capability to develop the *Atlas* ICBM, Schriever and Gardner hired R-W to serve as the technical direction contractor, an adviser to air force officers, and a technical watchdog over the contractors.

Feeling bogged down in "Wright Field procedures," external approvals, and funding difficulties, Schriever and Gardner appealed to President Eisenhower to break the logjam. The president complied, and so Schriever, armed with a presidential directive, hand-picked a committee to develop procedures that gave him the authority to acquire the services he needed from the air force without having to answer to the air force. The Gillette Procedures carved out a space in which Schriever, his officers, and scientific allies could craft their own development methods, largely separated from the air force's standard processes.

Under "concurrency," Schriever's complex of the WDD, the SAPO, and R-W created and adopted a number of methods to speed ICBM development. With funding a nonissue, these organizations and their contractors tossed aside standard regulations and developed alternate technical systems such as the *Titan* ICBM to ensure success. The air force's regular methods, based on academic-style disciplinary groups, no longer sufficed. Schriever broke away from dependence on Wright Field's technical groups and committees, but in the first years of ICBM development, he merely substituted his own officers and contractors, unencumbered by paperwork. The WDD, the SAPO, and R-W recreated an ICBM-oriented Wright Field on the West Coast, albeit without the years of history and bureaucracy.

The proof of their efforts would come when ICBM testing began in the late 1950s. As long as the Cold War remained hot and his scientific friends delivered technical success, Schriever could sustain concurrency. Unfortunately, tests would show that these new wonder weapons had major problems. Under these circumstances, politicians and managers would rein in the rapidly moving ICBM programs, replacing Schriever's all-out concurrency with a new, centralized bureaucracy that incorporated some of the key lessons of ICBM development.

THREE

From Concurrency to Systems Management

> We have found that concurrency is as unforgiving to inept management principles as a high performance aircraft is to pilot error. In fact, it requires *MORE* formality, not *LESS*.
>
> —Lieutenant Colonel Benjamin Bellis, 1962

By 1955, Bernard Schriever's Western Development Division (WDD), in conjunction with the Special Aircraft Projects Office (SAPO) and Ramo-Wooldridge Corporation (R-W), had implemented concurrency to rapidly move intercontinental ballistic missiles (ICBMs) from development into testing. As tests unfolded in 1956 and 1957, Schriever's officers and contractors found, much to their consternation, that *Atlas* failed at an alarmingly high rate. In the rush to push ICBMs into service, Schriever had created an organization that was remarkably informal and flexible but whose disregard of regular procedures also cut out many essential functions of the air force's bureaucracy. Many of these techniques had been put into place to ensure that there was communication among technical, financial, legal, and operational personnel. Focusing explicitly on the technical issues, Schriever's officers and contractors let other concerns fall to the wayside. Problems with financing and scheduling were compounded by technical problems endemic to radical new technologies.

To fend off criticism, Schriever's organization had to improve the reliability of the complex weapons and better predict and control costs. This required more formal engineering and management practices. Engineers made missiles more dependable through exhaustive testing, component tracking, and configuration control. Managers improved cost prediction and control using new tools like the Program Evaluation and Review Technique (PERT) and new procedures such as phased planning. The end result was systems management, a means to create new technologies rapidly but also to plan and control the excesses of concurrency. The new methods slowed development but increased reliability and cost predictability of air force technology programs.

While justifiable under the perceived national emergency in the 1950s, the huge costs of concurrency could not be sustained forever. To achieve cost control, Schriever and his cohorts adopted centralized, formal management techniques. Inherent in this shift was a slowdown in the pace of technological innovation, imposed by managerial checkpoints. Replacing a rapidly paced world of novel wonder weapons promoted by military officers and scientists was a more sedate world of dependable weapons and predictable administration offered by engineers and managers. Consistent with Secretary of Defense Robert McNamara's determination to centralize control and authority for weapons development, Schriever's modified techniques became the basis for the new Air Force Systems Command and by 1965 the heart of the Department of Defense's (DOD's) development processes.[1]

Systems Engineering and Black Saturdays

Systems management included techniques to improve engineering and reliability as well as methods for managers to coordinate and control large-scale development. The formal engineering methods ultimately used by Schriever's organization, known as systems engineering, derived largely from military programs in World War II and the early Cold War. Schriever's group would expand upon these ideas and ensure that they were adopted throughout the aerospace industry

Although historians have yet to determine all of the originators of systems engineering, many of them were involved with the military. In addition, systems engineering's proponents almost all had connections with one of two major technological universities in the United States: the California Institute of Technology or the Massachusetts Institute of Technology (MIT). At Caltech, most early systems proponents received their education under the tutelage of famed aerodynamicist and first head of the air force's Scientific

Advisory Board, Theodore von Kármán. MIT's systems approaches stemmed from the institute's direction of the Radiation Laboratory and other military projects during World War II.[2]

One the primary sources of systems engineering was the organizational culture of American Telephone and Telegraph (AT&T). Bell Telephone Laboratories, perhaps the single largest group of researchers in the United States outside of academia, performed research and development (R&D) for AT&T. Bell Labs researchers typically assigned hardware prototype manufacturing to Western Electric, AT&T's manufacturing arm. Because of the large size of the corporation and the multiplicity of projects, Bell Labs and Western Electric developed formal specifications and paperwork to handle the relationship between Bell Labs researchers and Western Electric engineers and manufacturing workers. In their relationships with outside contractors and the U.S. government, Bell Labs and Western Electric personnel found it natural to use these same formal methods. In this structured arrangement coordinating researchers and manufacturers was the kernel of systems engineering. Donald Quarles, who headed Bell Labs for a time and later became the assistant secretary of defense, was familiar and comfortable with Bell Labs' ideas about systems engineering. Mervin Kelly, who also headed Bell Labs, became an influential adviser to the air force on many systems.[3]

MIT became involved with Bell Labs and with systems engineering in part through the Radiation Laboratory's development of fire control systems during World War II. One major protagonist was physicist Ivan Getting, who worked on an MIT liaison committee that coordinated the integration of a Radiation Laboratory tracking radar to a Bell Labs gun director on the SCR-584 Fire Control System. He soon realized that because of electrical noise, the two components working together behaved differently than the two components alone. Getting had to analyze the behavior of the entire system, not just its components. Because of various wartime exigencies, Getting coordinated the efforts of General Electric, Chrysler, and Westinghouse to manufacture the system, acting as the de facto systems integrator and engineer for the project.[4]

Learning from this, in 1945 Getting made himself the liaison between the Radiation Laboratory and the navy's Bureau of Ordnance for the navy's Mark 56 project. He assigned the Radiation Laboratory as the system integrator for the project. The laboratory made all technical information available to General Electric and the navy, checked and criticized designs, sent representatives to conferences, reported to the Bureau of Ordnance on progress, participated in and established procedures for prototype, preproduction, and acceptance testing, and assisted in training programs. To accomplish these functions, Getting arranged for the Radiation Laboratory to receive copies of navy and contractor correspondence, drawings, and specifications; to be notified of significant tests and conferences; to examine production designs or models; to have access to contractors and their engineers; and to inspect equipment.[5] These arrangements established the formal function of system integration.[6] From Getting's position as a member of the air force's Scientific Advisory Board and as technical director for Air Defense Command, and from weapon systems engineering courses taught at MIT, the idea of systems engineering spread throughout the air force.[7]

The 1949 Ridenour Report that led to the founding of Air Research and Development Command (ARDC) noted, "The role of systems engineering should be substantially strengthened, and systems projects should be attacked on a 'task force' basis by teams of systems and component specialists organized on a semi-permanent basis." Transferring authority from the component engineers at Wright Field, the report recommended that the project officers and engineers who integrated components be given substantially more authority and autonomy.[8] Implementing the idea took a good deal of education and exhortation, along with new regulations. Maj. General Donald Putt, a protégé of Caltech's von Kármán, became commanding officer of Wright Air Development Center in 1952. He admonished the laboratory chiefs, "Somebody has to be captain of the team, and decide what must be compromised and why. And that responsibility we have placed on the project offices."[9] Engineering personnel in the project office acted as systems engineers, with the responsibility for the integration of technologies into the weapon system, whether aircraft or missiles.

Systems engineering also played a prominent role at Hughes Aircraft Company, where Simon Ramo had assembled a skilled team of scientists and engineers to develop electronic gear for military aircraft and the innovative *Falcon* guided missile. The *Falcon* differed from contemporary air-to-air missiles in that it used sophisticated electronics to guide the missile to its target and hit it. Other missiles typically placed a large warhead near an enemy aircraft, then detonated it nearby using proximity fuzes. These required substantial amounts of explosives and hence also a big, heavy missile to carry them. Ramo and Dean Wooldridge instead used what they called the systems approach to determine a more optimal design for air-to-air combat.[10]

Like MIT's Getting, Ramo formulated his notions of systems engineering through work on complex military projects. Although he had worked at General Electric, where a number of organizations had the word "system" in them, his work on various components did not stimulate any interest in the processes of engineering. Moving to Hughes Aircraft, and soon heading his own organization devoted to military electronics and missiles, Ramo began to think more seriously about the processes common to Hughes's varied tasks. Wondering how best to formulate and pass on the expertise necessary to address the complexities of missiles and electronic systems, Ramo began to promote the idea of an academic discipline of systems engineering. However, his first opportunities to pass along these ideas came not through publication but through his involvement with Schriever's ICBM program.[11]

Ramo's company came into being as a result of a meeting between Ramo and Secretary of Defense Charles Wilson in 1953. At that meeting, Wilson expressed displeasure that the eccentric Howard Hughes had captured a near monopoly on aircraft and missile electronics through Ramo's group. Wilson informed Ramo that he intended to "break this monopoly" and would support Ramo if he separated from Hughes. This catalyzed Ramo and Wooldridge's decision to form their own company, a decision soon rewarded when Deputy Secretary of the Air Force Trevor Gardner awarded them a contract to support John von Neumann's "Teapot Committee" (chapter 2). Gardner was an old friend of Ramo's, but despite this, Ramo and Wooldridge did not really want the ICBM systems engineering job because they correctly perceived that this would hinder their efforts to land lucrative hardware contracts. When the air force informed them in early 1954 that they would not acquire any air force contracts unless they took the ICBM systems engineering job, R-W accepted the contract from Schriever's group.[12]

Schriever fostered close working relationships between R-W, the WDD of ARDC, and the SAPO of Air Materiel Command (AMC). Schriever and Ramo agreed that R-W personnel should be placed in offices adjacent to those of

Brigadier General Bernard Schriever and Dr. Simon Ramo at a building dedication at the Inglewood complex in 1956. Courtesy John Lonnquest.

their WDD counterparts. For example, the office of Schriever's technical director, Charles Terhune, was next to that of the R-W technical director, Louis Dunn. At the highest level, Schriever and Ramo were in frequent contact.[13]

Despite the close contact, the function of R-W personnel was not clear to Schriever's group as late as April 1955. Schriever directed Ramo to assemble a briefing to describe for his officers and contractors the processes and tasks that R-W performed.[14] This briefing was one of the earliest descriptions of systems engineering. R-W formed its Guided Missile Research Division (GMRD) in 1954 to handle the technical aspects of the ICBM programs. With Ramo heading the division and Louis Dunn, the former Jet Propulsion Laboratory (JPL) director, as technical deputy, the GMRD in April 1955 had five departments: Aeronautics R&D, Electronics R&D, Systems Engineering, Flight Test, and Project Control. While the Aeronautics and Electronics departments concentrated on subsystems and components, the Systems Engineering, Flight Test, and Project Control departments performed the bulk of ICBM integration tasks.[15]

Technical direction of contractors took place through monthly formal meetings as well as numerous informal meetings. R-W Project Control personnel chaired the formal meetings, set the agenda, recorded minutes, and presented current schedules and decisions. Based on the results of these meetings, the Project Control Department issued Technical Directives, work statements, and contract changes. WDD officers reviewed Technical Directives, along with changes to work statements. They then submitted work statements and contract changes to the SAPO, whose officers then issued contractual changes and approved work statement modifications. Informal meetings were for "information only," and WDD and R-W personnel coordinated this information as necessary. The Project Control Department handled official plans, schedules, work statements, cost estimates, and contract changes.[16]

Engineers in R-W's Systems Engineering Department analyzed major design interactions, studied electrical and structural compatibility between subsystems and contractors, and issued top-level requirements. One good example was the nose cone trade study that cut *Atlas*'s mass in half. Another was an assessment of the Martin Company's trajectory analysis. Department members found that Martin's trajectory was less than optimal; by modifying it, R-W engineers increased the *Titan*'s operational range by 600 miles, the equivalent of saving 10% of its mass. R-W systems engineers performed experimental work in the laboratory when they needed more information, analyzed intelligence data on Soviet tests, and programmed early missiles. As noted by one critic of R-W, the engineers often double-checked contractors to avoid "errors, mistakes, and failures."[17]

By October 1956, the WDD and R-W came to a legal agreement about what systems engineering entailed. The agreement defined systems engineering in terms of three functions:

1. The solution of interface problems among all weapon system subsystems to insure technical and schedule compatibility of the systems as a whole.
2. The surveillance over detailed subsystem and over-all weapon design to meet Air Force required objectives.
3. The establishment and revision of program milestones and schedules, and

> monitoring of contractor progress in maintaining schedules, consistent with sound technical judgment and rapid advancement of the state of the art.[18]

From 1953 through 1957, R-W's role grew dramatically. Starting with documentation of the Teapot Committee's deliberations, R-W acquired a contract with Schriever's new organization to perform long-range studies of ICBMs, to assess new technologies, and to help the WDD set up and operate its new facilities. Its funding grew from $25,494 through June 1954, to $833,608 from July 1954 through June 1955, and to $10,095,545 from July 1955 through June 1956. As R-W's competence grew, Schriever expanded its role. R-W double-checked contractors' work; controlled specifications, schedules, and other paperwork; and surveyed the technical horizon for new technological solutions. As Schriever himself later admitted, R-W became for the WDD what Wright Field and its component engineers were for aircraft development. For the first few years of expansion, R-W's services were indispensable to the WDD; they cut program costs and improved ballistic missile performance.[19]

Along with systems engineering, Schriever initiated other methods to manage the program. As he well knew, the system approach required planning for the entire weapon life cycle from the start of the program. One of Schriever's first actions was to establish a centralized planning and control facility to facilitate application of this idea. The WDD established its own local and long-distance telephone services, including encrypted links for classified information and teletype facilities.

In the fall of 1954 Schriever and his staff developed a management control system. Every month, they required the air force, R-W, and associate contractors to fill out standardized status report forms. One of Schriever's officers controlled and updated the master schedules, placed on the walls of a guarded program control room. This room was both a place where managers could quickly assess the "official" status of the program and a place where Schriever and his deputies showed the program status and innovative management to visitors.[20]

A primary benefit of the management control system was the process of preparing the weekly and monthly status reports. Report preparation required that managers collect and verify data, identify problems, and make recommendations about how to resolve them. Schriever instituted monthly "Black Saturdays" for project officers to report difficulties. At these meetings, Schriever and his top R-W and military staff reviewed the entire program and assigned responsibility for resolving all problems to individuals there. These meetings endeavored to bring problems forward instead of sweeping them under the rug. As Schriever put it, "The successes and failures of all the departments get a good airing."[21]

While Black Saturdays brought some order to the technical aspects of the program, the Procurement Staff Division of the Ballistic Missiles Office at air force headquarters had to cope with the legal and financial mess created by Schriever's disregard for standard processes. The financial officers insisted that "the technical directives [be] covered by cost estimates" because annual funding from the DOD was insufficient to cover rising costs. Schriever fought these regulations as "examples of the 'law's delay'" but had to give in. In November 1956 he agreed to submit cost estimates, leading to new procedures in February 1957. To ensure that R-W and the other contractors documented technical directives, the Guidance Branch of the WDD in October 1956 "began holding a contract administration meeting immediately after each technical directive meeting." By January 1957 the Procurement Staff Division extended the practice to all technical direction meetings.[22]

With these new procedures to coordinate the legal and financial aspects of ICBMs, the air force could map out the ramifications of the various changes to the ICBM programs. Although this allowed for a modicum of order across the air force, only upcoming missile tests could determine whether the *Atlas* and the *Titan* would fly.

Testing Concurrency

Whatever preferences Schriever and his team had for rapid development, they insisted upon flight tests to detect technical problems. ICBMs were extremely complex, and some failures during initial testing were inevitable. Testing would uncover many problems as it began in late 1956 and 1957.

Missile testing differed a great deal from aircraft testing, primarily because each unpiloted missile flew only once. For aircraft, the air force used the Unsatisfactory Report System, whereby test pilots, crew members, and maintenance personnel reported problems, which were then relayed to Wright Field engineers for analysis and resolution. The problem with missiles was the lack of pilots, crew members, and maintenance personnel during development testing. Instead, manufacturers worked with the air force to run tests and analyze results.[23]

Because each ICBM disintegrated upon completion of its test flight, flight tests needed to be minimized and preflight ground testing maximized. The high cost of ICBM flight tests made simulation a cost-effective option, along with the use of "captive tests," where engineers tied the rocket onto the launch pad before it was fired. R-W engineers estimated that for ICBMs to achieve a 50% success rate in wartime, they should achieve 90% flight success in ideal testing conditions. With the limited number of flight tests, this could not be statistically proven. Instead, R-W thoroughly checked and tested all components and subsystems prior to missile assembly, reserving flight tests for observing interactions between subsystems and studying overall performance. Initial flight tests started with only the airframe, propulsion, and autopilot. Upon successful test completion, engineers then added more subsystems for each test until the entire missile had been examined.[24]

By 1955, each of the military services recognized that rocket reliability was a problem, with ARDC sponsoring a special symposium on the subject.[25] Statistics showed that two-thirds of missile failures were due to electronic components such as vacuum tubes, wires, and relays. Electromagnetic interference and radio signals caused a significant number of failures, and about 20% of the problems were mechanical, dominated by hydraulic leaks.[26]

Atlas's test program proved no different. The first two *Atlas A* tests in mid-1957 ended with engine failures, but the third succeeded, leading eventually to a record of three successes and five failures for the *Atlas A* test series. Similar statistics marked the *Atlas B* and *C* series tests between July 1958 and August 1959. For *Atlas D*, the first missiles in the operational configuration, reliability improved to 68%. Of the thirteen failures in the *Atlas D* series, four were caused by personnel errors, five were random part failures, two were due to engine problems, and two were design flaws.[27]

Solving missile reliability problems proved to be difficult. Two 1960 accidents dramatized the problems. In March an *Atlas* exploded, destroying its test facilities at Vandenberg Air Force Base on the California coast. Then, in December, the first *Titan I* test vehicle blew up along with its test facilities at Vandenberg. Both explosions occurred during liquid propellant loading, a fact that further spurred development of the solid propellant–based *Minuteman* missile. With missile reliability hovering in the 50% range for *Atlas* and around 66% for *Titan,* concerns increased both inside and outside the air force.[28]

While the air force officially told Congress that missile reliability approached 80%, knowledgeable insiders knew otherwise. One of Schriever's deputies, Col. Ray Soper, called the 80% figure "optimistically inaccurate" and estimated the true reliability at 56% in April 1960.[29] That same month, Brig. Gen. Charles Terhune, who had been Schriever's technical director through the 1950s, entertained serious doubts:

> The fact remains that the equipment has not been exercised, that the reliability is not as high as it should be, and that in all good conscience I doubt seriously if we can sit still and let this equipment represent a true deterrence for the American public when we know that it has shortcomings. In the aircraft program these shortcomings are gradually recognized through many flights and much training and are eliminated if for no other reason, by the motivation of the crews to keep alive but no such reason or motivation exists in the missile area. In fact, there is even a tendency to leave it alone so it won't blow up.[30]

ICBM reliability problems drew air force and congressional investigations. An air force board with representatives from ARDC, AMC, and Strategic Air Command reported in November 1960, blaming inadequate testing and training as well as insufficient configuration and quality control. It recommended additional testing and process upgrades through an improvement program. After the dramatic *Titan* explosion the next month, the secretary of defense requested an investigation by the Weapon Systems Evaluation Group within the Office of the Secretary of Defense. A parallel study by the director of Defense Research and Engineering criticized rushed testing schedules. In the spring of 1961, the Senate Preparedness Investigating Subcommittee held hearings on the issue. The members concluded that testing schedules were too optimistic. With technical troubles continuing, its own officers concerned, and congressional pressure, Schriever's group had to make ICBMs operationally reliable. To do so, the air force and R-W created new organizational processes to find problems and ensure high quality.[31]

Solving ICBM technical problems required rigorous processes of testing, inspection, and quality control. These required tighter management and improved engineering control. One factor that inadvertently helped was a temporary slowdown in funding between July 1956 and October 1957. Imposed by the Eisenhower administration as an economy measure, the funding reduction slowed development from "earliest possible" deployment (as had been originally planned) to "earliest practicable" deployment. As noted by one historian, this forced a delay in management decisions regarding key technical questions related to missile hardware configurations, basing, and deployment. This, in turn, allowed more time to define the final products.[32]

Reliability problems were the most immediate concern, and AMC officers began by collecting failure statistics, requiring *Atlas* contractor General Dynamics to begin collecting logistics data, including component failure statistics, in late 1955. In 1957 AMC extended this practice to other contractors, and it later placed these data in a new, centralized Electrical Data Processing Center.[33] R-W scientists and engineers statistically rationed a certain amount of "unreliability" to each vehicle element, backing the allocations with empirical data. They then apportioned the required reliability levels as component specifications.[34]

Starting on *Atlas D*, Space Technology Laboratories (STL)—the successor to R-W's GMRD—scientists and engineers began the Search for Critical Weaknesses Program, in which environmental tests were run to stress components "until failure was attained." The scientists and engineers ran a series of captive tests, holding down the missile while firing the engines. All components underwent a series of tests to check environment tolerance (tolerance for temperature, humidity, etc.), vibration tolerance, component functions, and interactions among assembled components. These required the development of new equipment such as vacuum chambers and vibration tables. By 1959, the Atlas program also included tests to verify operational procedures and training. STL personnel created a failure reporting system to classify failures and analyze them using the central database.[35]

Environmental testing, such as acoustic vibration and thermal vacuum tests, detected component problems. The failure reporting system also helped

Atlas D launch, October 1960. *Atlas* reliability began to improve with the D series. Courtesy John Lonnquest.

identify common weaknesses of components. Other new processes, such as the Search for Critical Weaknesses Program, looked for problems with components and for troublesome interactions. These processes identified the symptoms but did not directly address the causes of problems. For example, some component failures were caused by a mismatch between the vehicle flown and the design drawings. Solving problems, as opposed to simply identifying them, required the implementation of additional social and technical processes. Engineers and managers created the new social processes required on the Minuteman project, and from there they spread far beyond the air force.

The Creation of Configuration Management

The Minuteman project was the critical turning point for air force ICBM programs. First, its use of solid propellants instead of troublesome liquids greatly simplified and decreased the dangers of ICBM launch operations. Second, the Minuteman assembly and test contractor, Boeing, brought to the Inglewood complex a new management technique that would become the centerpiece of the air force's R&D management process: configuration management. The combined effect of solid propellants and configuration management was a dramatic improvement in ICBM reliability and cost predictability.

By 1957, a solid-propellant alternative to troublesome liquid-fueled ballistic missiles such as *Atlas* and *Titan* became feasible. Col. Ed Hall, Schriever's propulsion manager, had, along with R-W and a number of contractors, been studying solid propellant technology for some time and had evidence to show that it could be developed for large-scale ICBMs. Hall pointed out to Technical Director Charles Terhune that solid-propelled rockets did not involve costly, time-consuming, and dangerous liquid-propellant loading procedures. Although liquid-propelled rockets had higher performance, it took several hours to prepare them for launch. Solids, on the other hand, could be launched within seconds; once loaded with propellants and placed in their launch configuration, they were ready to go with the push of a button. Hall now had evidence to show that solids, which heretofore could perform adequately with only a small size, could now be manufactured and perform adequately on a much larger scale, making solid-propellant ICBMs feasible.[36] Solid propellants eliminated dangerous liquid-propellant loading operations that destroyed launch pads and killed workers. This in itself was a tremendous advantage. However, the use of solid propellants did nothing to fix subtle problems associated with unintended component interactions resulting from poor designs, or worse, resulting from flight hardware that did not match anyone's design.

On *Atlas* and *Titan* test flights, engineers found that a number of test failures resulted from mismatches between the missile's design and the hardware configuration of the missile on the launch pad. In the rush to fix problems, the launch organization, contractors, or air force had made modifications to missiles without documenting those modifications. To fix this problem, STL personnel and air force officers developed a reporting procedure known as configuration control to track and connect missile design changes to missile hardware changes. Because these often involved manufacturing and launch processes, configuration control soon controlled process changes as well.[37]

While inspired by problems endemic to ballistic missiles, configuration control drew from the Boeing Company's aircraft programs. The air force learned about configuration control through the Minuteman project, where Boeing was the assembly and test contractor. Boeing's quality assurance procedures used five control tools:

1. formal systems for recording technical requirements
2. a product numbering and nomenclature system for each deliverable contract item
3. a system of control documents with space for added data on quantities, schedules, procedures, and so forth
4. a change-processing system
5. an integrated records system[38]

In addition, Boeing's "change board" ensured that all affected departments reviewed any engineering or manufacturing change and committed appropriate resources to effect it. The air force soon saw the importance of this process innovation and made it into a critical new management process, with its own organization and staff.[39]

Boeing's processes supplanted the concept of the "design freeze." The design freeze was an important milestone in aircraft development, the point when engineers stopped making design changes so that hardware could be built to that design. Once the design was frozen, engineers or operators could make design changes only by submitting a formal change request. Engineers then made sure that corresponding changes were made to the hardware and the production facilities.

Ballistic Missile Division (BMD, successor to the WDD) officers and STL engineers used configuration control to coordinate changes and ensure the compatibility of designs and hardware. The key to configuration control was the creation of a formal change board with representatives from all organizations, along with a formal system of paperwork that linked specifications, designs, hardware, and processes. Although initially linking design drawings to hardware, BMD officers and R-W engineers soon realized that by expanding configuration control to include specifications and procedures, they could control the entire development process.

Through configuration control, systems engineers linked specifications to designs, designs to hardware, and hardware to operational and testing procedures. Engineers brought proposed changes to the configuration control board. Air force officers soon linked configuration control to contracts, tying engineering changes to contract changes. The air force established configuration control in the fall of 1959 on Minuteman; soon, configuration control had been extended to its other space and missile projects.[40] Officers and managers vigorously promoted configuration control because of its utility in linking engineering, management, and contracts.

By the early 1960s, the coordinating role of STL and The Aerospace Corporation[41] had evolved into a procedure called systems requirements analysis, in which technology development was managed through the control of requirements. For example, at the highest level a requirement would be written to develop a ballistic missile system to deliver a one-megaton payload over 5,000 miles with an accuracy of 1 mile. Systems engineers divided this requirement into at least three statements at the next level. These three would then be broken down into numerous requirements to create hardware components, operating procedures, and so on. Major programs involved thousands of requirements, corresponding to thousands of components and procedures. Systems requirements analysis made the design traceable to requirements of increasing specificity.[42]

Detailed requirements analysis, and more importantly, configuration control, found a powerful advocate in Col. Samuel C. Phillips, who in 1959 replaced Col. Ed Hall as the manager of Minuteman.[43] Phillips, who graduated in 1942 with a bachelor's degree in electrical engineering from the University of Wyoming, had steadily worked his way up the air force's hierarchy as a skilled technical manager. After serving as a pilot in Europe in World War II, he started in 1950 as a project engineer at Wright-Patterson Air Force Base. Through the 1950s, he held an assortment of positions, including electronics officer for atomic weapons tests at Eniwetok Atoll, chief of operations at the Armament Laboratory at Wright-Patterson, project officer for the *B-52* bomber, chief of the bombardment aircraft division, chief of the fighter missiles and drones division, and eventually, logistics chief and materiel director of Strategic Air Command in England. Phillips was quiet, forceful, and tactful, and he brought the tools developed by Schriever's team to Minuteman.[44]

In 1959, managers at Boeing, or possibly Phillips himself, realized that by a simple extension of configuration control, they could gain financial as well as technical control over the project.[45] The idea was simple. All a project manager had to do was compel engineers to give cost and schedule estimates along with any technical change. If the engineer did not give the information, the project manager rejected the change. With this added information, the project manager could predict and revise the project's cost profile, along with its schedule. This also allowed the manager to track the performance of each engineer or group of engineers—he could now hold them accountable to their own estimates. Managers tied the process to specific design configurations and eventually to the hardware itself. In addition, procurement officials and industry managers could write contracts against specific design configurations and negotiate cost changes based upon each approved change. Phillips and others transformed configuration control into "configuration management," a critical managerial tool to control the entire development process. Others soon recognized its utility, leading in 1962 to the creation of general regulations and guidelines for configuration management.[46]

Through configuration management, and also because of its solid-propelled rocket design, *Minuteman* boasted an enviable development record, coming in on cost and on schedule. Because of *Minuteman*'s much better reliability and launch-on-demand capability, the air force soon phased out most liquid-propellant missiles as weapons. Higher performance made liquid-propellant missiles excellent satellite launchers, and they and their descendants performed well in this role. In their capacity as launchers, the *Atlas*, *Titan*, and *Thor-Delta* vehicles attained reliability exceeding 90% from the mid-1960s through the 1990s.[47]

Configuration management—along with further attention to quality through inspections, training, and associated documentation—became an organizational pillar of the air force's management system. Its importance can hardly be overstated. Managers from the turn of the century through the 1950s had searched for ways to predict R&D costs and to control scientists and engineers. Configuration management achieved this control on development projects, as it allowed accountants and lawyers to tie technical modifications to contract modifications, including costs. Configuration management enabled government to control industry. However, the government officials who wielded the authority had make clear distinctions between those doing the controlling and those being controlled. To do this, they would have to modify the anomalous position of R-W.

The Formation of The Aerospace Corporation

From the beginning of the WDD, aircraft industry leaders complained bitterly about R-W's insider position. They believed that the ideal approach to weapons development was for the air force to let prime contracts to a single integration contractor, a position supported by the air force's own regulations. These stated that the air force should hire a single prime contractor to develop, integrate, and test a weapon system, unless no company was qualified to perform the task. In this case, the air force itself could act as prime contractor. The latter position was Schriever's justification for his approach to the ICBM program, with the important modification that the air force would instead hire a third party to direct technical coordination of the integration task. Industry leaders also pointed out that in R-W, the air force was creating a new, powerful competitor with close ties to air force planning and a concomitant edge in bidding.[48]

Normally, R-W should have been controlled by the air force in the way that any other contractor would have been. However, the air force had hired R-W to act as the air force's technical assistant for ICBM development, in which position R-W personnel acted with virtually the same authority as the government. In 1954, Assistant Secretary of Defense for Research and Development Donald Quarles, formerly of Bell Labs, had insisted that R-W personnel be given "line" responsibility, with full authority to direct contractors, instead of "staff" status, where they would merely be advisers. This mirrored his experience at Bell Labs, which acted as the technical direction authority to AT&T's manufacturing arm, Western Electric. Bell Labs also performed this role with other contractors, sometimes on behalf of the government on high-priority military programs. This powerful position required that AT&T acquire sensitive data from other companies. As a regulated monopoly, AT&T could legitimately act in this capacity, as it essentially had no competitors.[49]

Caltech's JPL and MIT's Radiation Laboratory also acted as technical direction groups for the government, but these academic nonprofit institutions were little threat to industry. However, R-W was neither a nonprofit institution nor a regulated monopoly, and in fact it competed for other projects against the same companies that it monitored on the ICBM program. Existing aircraft firms vigorously campaigned against the air force's unusual relationship with the upstart company.

To protect his organization from criticism, Schriever enforced a hardware ban on R-W to keep it from acquiring lucrative hardware contracts on any programs in which it was the technical direction contractor. R-W "walled off" the technical direction work of STL from the rest of the company. Continuing concerns led R-W to establish a physically separate location for its headquarters—in Canoga Park, California. These measures did not satisfy industrial leaders, who continued to lobby against the company.[50]

Despite the clamor and the ICBM hardware ban, neither Ramo nor Wooldridge believed that R-W could grow without manufacturing capabilities. They grasped every opportunity to expand manufacturing by aggressively pursuing hardware production products and contracts outside the ballistic missile program. These included process control computers, semiconductors, and a variety of aircraft and air-breathing missile components. Aggressive pursuit of hardware contracts paid off, as R-W received permission to build ballistic missile hardware to test ablative nose cones built by General Electric. Strongly backed by Schriever's technical director, Col. Charles Terhune, STL then built the *Able 1* lunar probe launched in August 1958 and the *Pioneer 1* spacecraft launched by the National Aeronautics and Space Administration (NASA) in October 1958. These activities fomented even more severe industrial protests, as the hardware ban against R-W evaporated.[51]

Expansion on these and other ventures such as semiconductors stressed R-W's finances. Ramo and Wooldridge leaned on their original investor, Thompson Products, for cash to expand facilities and capital equipment, and the ensuing negotiations led to an agreement that resulted in the merger of the two companies effective October 31, 1958. The new combination, Thompson-Ramo-Wooldridge (TRW), became the aerospace giant that the older aircraft companies had feared.

TRW executives recognized the awkward position of STL in the new company. STL handled TRW's space business, including both the technical direction tasks for the air force and STL's budding space manufacturing businesses. Because of the air force connection, STL would always be vulnerable to charges of conflict of interest. To minimize this risk, TRW executives established STL as an independent subsidiary corporation with its own board of directors chaired by Jimmy Doolittle, a war hero with impeccable credentials and impressive ties to the air force and NASA. No TRW board member or senior manager sat on STL's board. TRW executives recognized that they might have to divest STL, and through this reorganization they were prepared to do so.[52]

Although TRW was prepared to divest STL, neither Schriever nor TRW really wanted this to happen. TRW enjoyed significant profits from STL, and Schriever wanted STL's experienced personnel directing the technical aspects of the air force's ICBM and space programs. However, STL's increasing involvement with space projects and hardware development fueled industry complaints, leading to congressional hearings in February and March 1959.

These hearings, chaired by Rep. Chet Holifield from California, featured vehement attacks against STL's "intimate and privileged position" with the air force and equally strong defenses by Schriever and by TRW executives Simon Ramo and Louis Dunn. It became clear even to Schriever that as long as TRW acquired competition-sensitive technical information from other aerospace firms through STL, the clamor would continue. A plan to sell STL to public investors fell through when Air Force Secretary Douglas vetoed it on the grounds that STL would remain a problem as long as private owners used STL to make a profit. The Holifield Committee's final report seconded this idea and urged that STL be converted into a nonprofit corporation like RAND and MITRE. Schriever reluctantly agreed, leading to the formation of The Aerospace Corporation on June 4, 1960.[53]

At Schriever's insistence, STL continued systems engineering and technical direction for the ballistic missile programs for the near future, but all others transferred to Aerospace. Dr. Ivan Getting became Aerospace's first president, and a number of STL personnel transferred to the new corporation. This ended the controversy about TRW's insider position with the air force, but as industry had feared, there was a powerful new competitor with which to contend. Aerospace became one of a growing breed of nonprofit corporations that served the air force and other military organizations.

Systems engineering, which required the coordination of all elements of the technical system, could be performed by a prime contractor for the system, by the air force itself, or by a nonprofit firm that had no interest in competition. The experience of R-W showed that a profit-making corporation could not act on behalf of the U.S. government to coordinate or control the efforts of its competitors. The function of systems engineering had to be contained within the government itself, a neutral third party hired by the government such as Aerospace or MIT, or a prime contractor. With this controversy settled, the air force could now standardize systems management as its primary R&D method across all of its divisions.[54]

Standardizing Systems Management

By 1959, ongoing deliberations at air force headquarters were under way regarding the applicability of Schriever's "Inglewood model" to the rest of the air force's development programs. A senior committee headed by the AMC commander, Gen. Samuel Anderson, agreed that the air force should adopt the methods used in Inglewood, with the planning and implementation of new projects on a systems, or "life cycle," basis. Planning for the entire system would occur up front, and project offices would have the authority to manage development, including funding authority. However, the committee split into three camps regarding the organization, advocating positions ranging from minor modifications to radical reorganization. In June 1960, the Air Staff selected the least ambitious plan, which did include installation of new regulations based on Schriever's organizational processes, to be used on all the air force's major development programs.[55]

The 375-series regulations for systems management originated with one of Schriever's officers, Col. Ben Bellis, who headed an effort to document the procedures developed in Inglewood. After a series of reviews, the new regulations for systems management appeared on August 31, 1960, and were applied to the air force's major projects for missiles, space, aeronautics, and electronics. Subsequently revised and extended, these regulations became the institutional backbone of the new, Inglewood-inspired R&D system.[56]

Under the new regulations, the system program director gained significant authority. The air force required that the program director create and gain approval of a single document known as the System Package Program. Each System Package Program provided information on cost, schedule, management, logistics, operations, training, and security.[57] The 375 regulations formally applied the ARDC-AMC project office concept across all air force major acquisition programs.

The more radical "Schriever Plan" to manage the air force's R&D had been shelved by Anderson's committee in 1959, but it gained new life in 1961 when Robert McNamara became secretary of defense. McNamara, trying to resolve the controversy over which service should gain the coveted military space mission, looked for evidence of managerial and organizational expertise to determine which service should lead space efforts. With several hints from the McNamara camp that the Schriever Plan would help the cause, Air Force Chief of Staff Thomas White approved it. Secretary of the Air Force Eugene Zuckert and McNamara signaled their pleasure by conferring all space research to the air force in March 1961.[58]

The Schriever Plan reallocated the procurement activities of AMC to a new organization that also included the development functions of ARDC. ARDC was abolished, its place taken by Air Force Systems Command (AFSC), which came into being on April 1, 1961. Schriever, appointed the first commander of AFSC, now managed all of the air force's major development programs in four divisions: the Ballistic Systems Division in San Bernardino, California; the Space Systems Division in El Segundo, California; the Aeronautical Systems Division in Dayton, Ohio; and the Electronics Systems Division in Lexington,

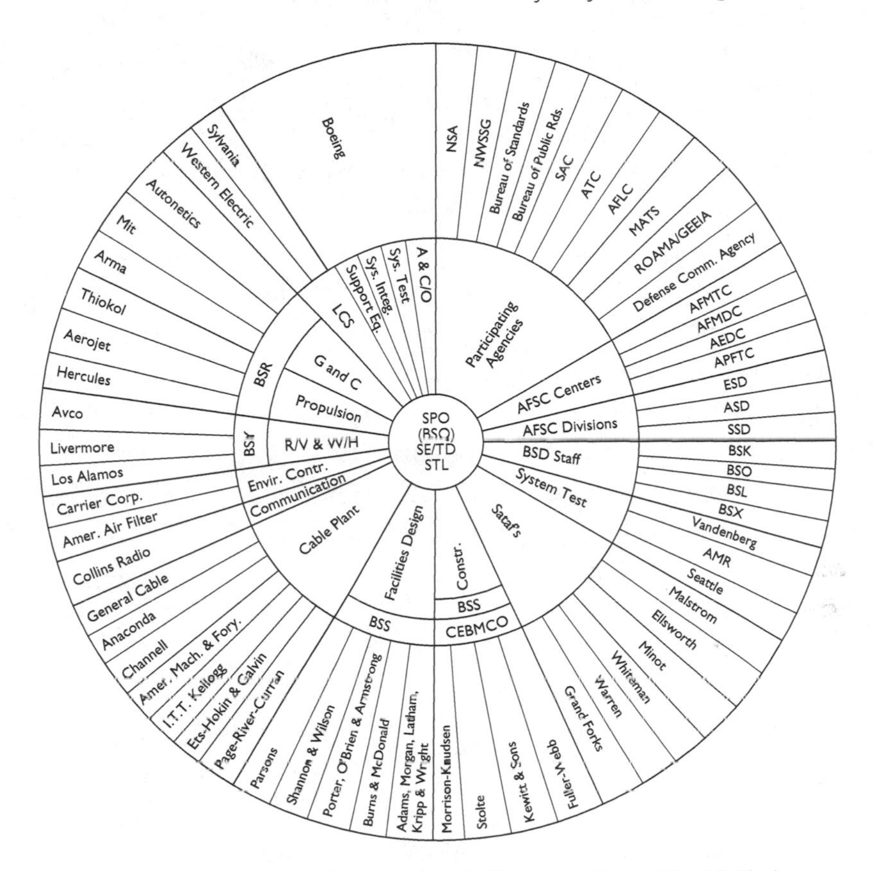

The air force's Ballistic Systems Division and Thompson-Ramo-Wooldridge's Space Technology Laboratory were in the center of a vast network of government and industry organizations, all of which learned aspects of systems management. "BSQ" represents the Ballistic Systems Division, and "SE/TD" stands for systems engineering and technical direction, the main function of STL. Courtesy Library of Congress.

Massachusetts. Ascending to command over all of the air force's large acquisition programs, Schriever's presence ensured the spread and enforcement of the 375 procedures.[59]

Standardization of R&D in AFSC went beyond the 375 regulations. By mid-1961, Schriever's organization molded status reporting into a highly sophisticated system, known as rainbow reporting because it presented each element of the system on pages of different colors in a small, brightly packaged booklet. Over the next few years, the rainbow reporting system evolved to include yearly and monthly milestone schedules, government and contractor financial data, contractor manpower data, reliability data, procurement data, engineering qualification data, and the so-called PRESTO procedures for problems needing immediate attention. They also specified acceptable formats and technologies for presentations to ensure commonality, helping the top-level managers to judge the programs on a consistent basis.[60]

With the establishment of AFSC, the Inglewood model of systems management, including configuration management, became the dominant model for large-scale programs. In April 1961, Schriever's authority and influence reached its apex, as he presided over all major development programs in the air force, using standardized methods of his own making.[61] What Schriever and others did not foresee was that just as the air force could use systems management to control contractors and its own officers, so too could the DOD use it to control the air force.

McNamara, Phased Planning, and Central Control

Within the DOD, the Office of the Secretary of Defense grew in power from 1947 to the mid-1960s. Over the years, the office progressively pulled critical decisions up the hierarchy, subordinating service interests and rivalries. Benefiting and exploiting this trend to the fullest was John F. Kennedy's appointee to the office, Robert McNamara.[62]

McNamara trained at the University of California, Berkeley, and taught business courses for a short time at Harvard before World War II. During the war, he performed statistical analyses for army logistics, determining the quantities of replacement parts needed based upon statistical assessments of combat and operations. After the war, he joined Ford Motor Company, tagged as one of the mathematically trained "whiz kids" that reformed Ford's disorganized finances and helped turn the company around. He rose quickly, eventually becoming president.[63]

Famous for his faith in centralized control implemented through quantitative measurement, McNamara took advantage of the authority granted to the Office of the Secretary of Defense by the Defense Reorganization Act of 1958. This act gave the secretary of defense the authority to withhold funding from the services and transfer assignments between the services. Upon his appointment to the office, in the spring of 1961 McNamara initiated a series of more than 100 studies known as McNamara's 100 trombones, or the 92 labors of Secretary McNamara. The services readily complied with this request, expecting the novice secretary to get bogged down in conflicting piles of recommendations.[64]

Without waiting for completion of the studies, McNamara also installed RAND chief economist Charles Hitch as the DOD comptroller. Given McNamara's background as a Ford financial manager and Hitch's qualifications as an economist, it was not surprising that they considered economic criteria to be foremost in making decisions for future weapon systems. Hitch's Program Planning and Budgeting System required that life cycle cost estimates be performed before deciding whether to develop a new weapon system. This agreed with the result of one of McNamara's studies—"Shortening Development Time and Reducing Development and Systems Cost"—which claimed that "reducing lead time and cost" should be given the same priority as improving performance. It deemphasized the relentless push to higher technical performance and required that feasibility and effectiveness studies calculate technical risks and cost-to-effectiveness ratios.[65]

Following up on this study, in September 1961 McNamara assigned the task of improving R&D management to John Rubel, the deputy director of defense research and engineering. Rubel established model programs whose methods could then be copied throughout all of the services, starting with the air force Agena, TFX fighter, Titan III, and medium-range ballistic missile programs. Rubel required a "Phase I" effort to develop a preliminary design. This would ensure "that the cost estimates for the subsequent development effort" were "based on a solid foundation."[66] The preliminary design effort would generate "a set of drawings and specifications and descriptive documents" to describe management methods, including schedules, milestones, tasks, objectives, and policies. Rubel had no reservations about forcing industrial contractors to organize and manage their projects in the way he wanted. If they wanted the job, they had to conform.[67]

He made clear in the request for proposals that go-ahead for Phase I did not constitute program approval. Previously, award of a preliminary design contract constituted de facto project approval for development and production. This was no longer true. Only the secretary of defense could approve a project, and he would not do so until completion of Phase I and a program review.[68] According to Rubel, "The fact that improved definition is required before larger-scale commitments are undertaken is neither surprising nor unique, although it is true that on most programs this definition phase has been less clearly identifiable because it has been stretched out in time and interwoven with other program activities such as development, model fabrication, testing and, in some cases, even production." Rubel did not believe that a program definition phase would slow high-priority programs. "In fact," he wrote, "our real progress should be accelerated as the result of obtaining a better focusing of our efforts."[69]

The phased approach brought several benefits to upper management. It promised better cost, schedule, and technical definition. If the contractor or agency did not provide appropriate information, management could cancel or modify the program. Organizations therefore made strenuous efforts to finalize a design and estimate program costs. The preliminary design phase provided management with a decision point before spending large sums of money, making projects easier to terminate and contractors easier to control.

By 1962, studies by Harvard and RAND economists had shown that DOD weapons projects had consistently large overruns and schedule slips, with missile programs having the worst record. The RAND study showed that for six missile projects, costs overran by more than a factor of four, with schedule slips greater than 50%. Other projects showed smaller slips, but all types averaged at least 70% cost overruns, and the average was more than 200% (triple the original cost estimates). The military was clearly vulnerable to criticism on cost issues, and McNamara efficiently exploited this weakness. His Program Planning and Budgeting System required that all of the services create five-year projections of programs and their costs, allocated not by specific services but rather across broad categories such as strategic offense or defense.[70]

Schriever sensed the change in national priorities and saw the impact of McNamara's reforms. Replacing "concurrency," "managerial reform" and "cost control" soon became the new watchwords. The immediate task facing Schriever in early 1962 was responding vigorously to the McNamara-Rubel initiatives, which he saw as cost control measures. In a February 1962 memorandum, Schriever stated that cost overruns arose from "any one or a combination of" factors, including deliberate underestimation, adherence to overly strict standards, too much optimism in estimating performance and schedules, vacillation or changes in program direction, and inadequate military or contractor management.[71]

One area that Schriever had to improve was cost estimation. His comptroller's office began by educating AFSC staff, instituting cost analysis training courses at the Air Force Institute of Technology in Dayton, Ohio. By February 1962, the first class of 25 students graduated from this course. AFSC also developed the Program Planning Report, which allowed for improved analysis of cost data with respect to technical and schedule progress. He also had AFSC adopt and modify the navy's new planning tool, PERT.[72]

Schriever developed other ways to improve AFSC's management capabilities. He established a Management Improvement Board, "made up of General Officers having the greatest experience in systems management matters ranging from funding, systems engineering, procurement and production, through research and development." Schriever had board members examine "the entire area of systems management methods to include those of the Industrial complex as well as those of the Air Force." He also reinstated the Air Force Industry Advisory Group, a Board of Visitors to improve working relationships with industry, and a program of "systems management program surveys." AFSC also collected "lessons learned" information from programs and broadcast this information through publications and industry symposia. Schriever also used this information to produce management goals for AFSC.[73]

AFSC also communicated systems management concepts through education. Examples included a system program management course at the Air Force Institute of Technology and the creation of a systems management newsletter within AFSC. The Air Force Institute of Technology course used case studies taught by experienced program managers such as Col. Samuel Phillips of the Minuteman program. These program managers taught about program planning and budgeting, the McNamara reforms, organizational roles in system development, systems engineering, configuration management and testing, system acquisition regulations, program management techniques, contracting approaches, and financial methods.[74]

By the mid-1960s, the combination of AFSC management initiatives and the McNamara reforms produced a mature form of systems management that is still used in the aerospace industry today. Earlier concepts and practices of

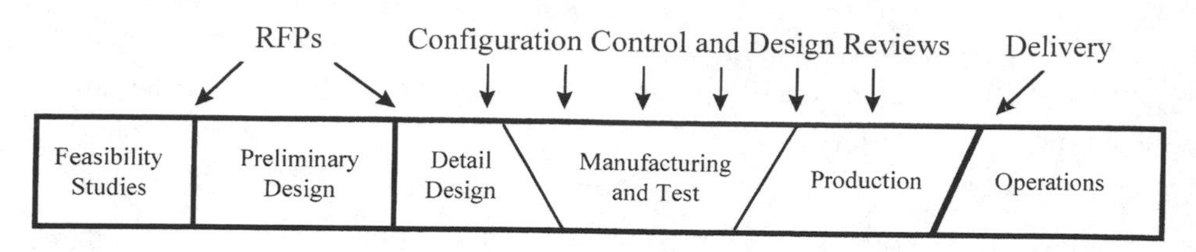

Systems management phases.

concurrency contributed the detailed planning and systems engineering coordination necessary to rapidly develop large-scale technologies. When ICBM failures became the primary concern, engineers added change control, quality control, and reliability to the mix. Finally, the cost concerns of the early 1960s—driven by rising ICBM costs, the Vietnam War, and social issues such as the civil rights movement—contributed phased planning and configuration management. Both new methods provided mechanisms to better predict costs.

McNamara, duly impressed with the procedures and reforms in Schriever's organization, used them—modified to include phased planning for central control—as the basis for the DOD's new regulations for the development of large-scale weapon systems. In 1965, the DOD enshrined phased planning and the systems concept as the cornerstone of its R&D regulations. Having already spread to NASA, these processes moved throughout the aerospace industry. Even when the processes were not explicitly used, industry accepted the assumptions and ideas encompassed in these regulations.[75]

Multiple Systems: Matrix Management

While air force officers funded government R&D, industry performed most of the work. Because of industry's dependence upon the military for R&D and production contracts, changes in the air force's organization and procedures had significant ramifications for industry. Industry managers grappled with government directives and technical problems, resulting in a variation of the military's model for technical organization: matrix management.

After World War II, aircraft companies shrank with the reduction in government contracts and observed the services' organizational and technological changes with interest. For missile programs and complex aircraft, aircraft companies built entire systems that included ground equipment, armament, and electronics. Aircraft companies began to reorganize their efforts around the complex new products. When the air force reorganized on a systems basis in 1952, aircraft contractors were well prepared. Each company, with a number of complex projects under way, had to reconcile the new project-based organization with their traditional discipline-based, functional structure.

The Martin Company developed one of the first project management organizations in the years 1952–53. William Bergen, an engineer who headed the pilotless aircraft group and the contract for the Naval Research Laboratory's *Viking* rocket in the late 1940s and early 1950s, was an early promoter of system management. As he wrote in a 1954 *Aviation Age* article, "Within the company we have created a number of miniature companies, each concerned with but a single project. The project manager exercises overall product control—in terms of an organization of all skills." Martin quickly implemented Bergen's innovation and expanded it to "cover all functions from design through manufacturing and distribution."[76] The Martin Company's systems approach included three elements: systems analysis to determine what to build, systems engineering to design it, and systems management to build it.[77]

Another example was McDonnell Aircraft Company's F-4 Phantom program. In the 1940s, McDonnell designed aircraft by committees staffed with engineers from the functional departments, with owner J. S. McDonnell arbitrating disputes. When the navy awarded the company a contract to develop the *F-4* in 1953, the company made young engineer David Lewis its first project manager. Lewis assigned three project engineers outside of the functional departments' jurisdiction to make design decisions, while he built the project organization and acquired resources.[78]

Project management involved the separation of engineers from their functional departments so that the engineers reported directly to a project manager whose sole task was to run the project. As projects grew, the number of managers and engineers also grew, but all reported to project managers, not functional department managers. As explained by a business school professor in 1962, "*The primary reason for project management organization is to achieve some measure of managerial unity,* in the same way that physical unity is achieved with the project."[79]

With numerous projects under way, military contractors faced the problem of developing several of them concurrently. The old line-and-staff organization no longer sufficed, as communication lines across functional departments became too long for effective coordination. H. F. Lanier, a project engineer for Goodyear Aircraft's Aerophysics Department, explained:

> The problem can perhaps be best illustrated by considering the difficulties of trying to fit a number of creative people into the precise and orderly line organization shown in [the "traditional line organization" figure]. Under this plan, all work is thoroughly organized and all assignments rigidly controlled. Each individual has a definite area to cover, definite data to work with, and a schedule to meet. He also has a boss who tells him what to do and subordinates whom he tells what to do. This organization once set up is soon limited to the creative output of a few men who lead. Any innovation is difficult to introduce because it requires detailed instruction at all levels.[80]

Lanier concluded that the long lines of communication needed breaking down. Ad hoc means did not suffice over the long term or for large projects: "The usual solution was to allow a great deal of 'co-ordination' and 'liaison' to be handled informally. Effectively, supervisors unleashed their men and gave the program general direction but let detailed instructions be formulated after the fact. The loose method has been reasonably successful. The next obvious step is to attempt to systematize the process."[81]

Often the first attempt at systematization was to form committees of functional supervisors. This did not work once meetings became too large or too frequent: "Usually the committee members are also line supervisors and hence can meet only for a fraction of the time required for efficient system development. In other words, actual development by a committee is employed most effectively on an occasional relatively huge problem. When large systems problems are the prime business, then a permanent fix must be made."[82]

For this purpose, Lanier explained, "The solution seems to be a committee of project or systems engineers—individuals trained to be jacks of all trades, and who are relieved of line responsibility for administering operating sections." Project management aligned engineers to the project but left undetermined their relationship to the rest of the organization. Lanier recognized that engineers had relationships to both the project and the remainder of the organization, where the engineers had to go when the project ended. Engineers

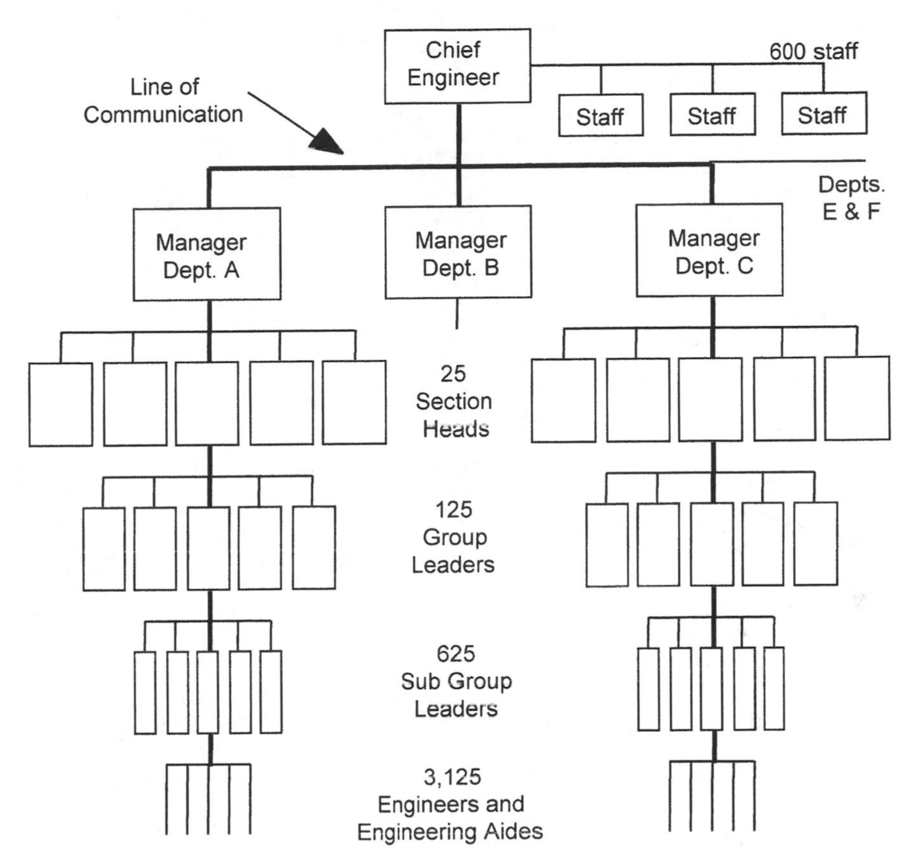

Traditional line organization and lines of communication. For complex projects, the communication lines become too long. Adapted from H. F. Lanier, "Organizing for Large Engineering Projects," *Machine Design* 28 (1956): 54.

therefore reported to both project and line management. Lanier called this dual reporting the project-line combination organization. The new organizational form had a "two-dimensional" or "matrix" form.[83]

The evolution of General Dynamics' Astronautics Division, responsible for the *Atlas* missile, typified the organizational changes brought on by the division's involvement in several military projects. For most of the 1950s, Convair ran Atlas as a single-project organization. Initially, Atlas (then known as Project MX-774) was directed "by a project engineer who was assigned a small team of designers and technical specialists plus an experimental shop for fabrication of the hardware." By 1954, one year after the acceleration of the Atlas

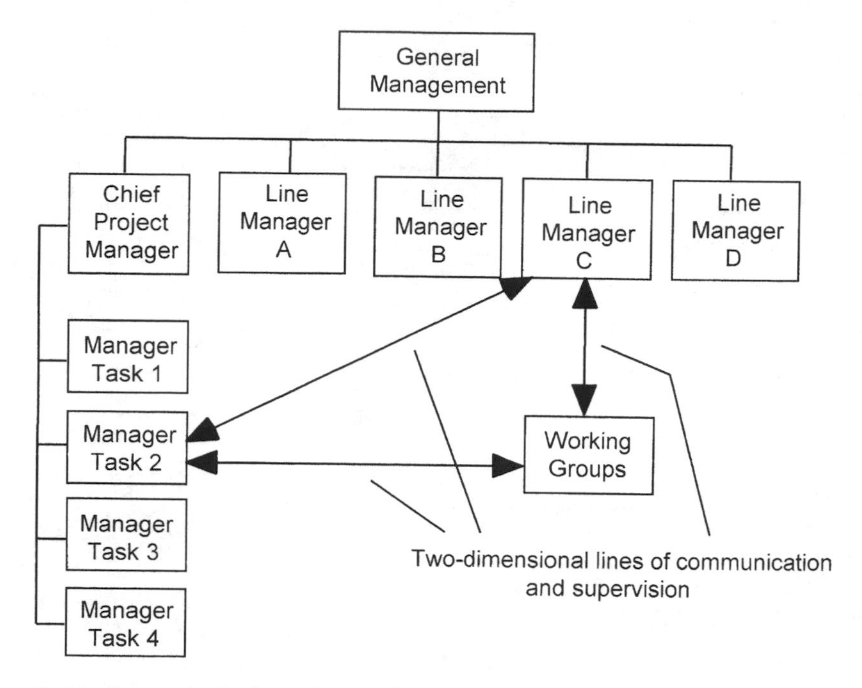

Project-line method of organization, later known as the matrix organization. Adapted from H. F. Lanier, "Organizing for Large Engineering Projects," *Machine Design* 28 (1956): 57.

project, Convair reorganized the project around the program office and had a force of 300 personnel, mostly engineers. In 1955, the company created the Astronautics Division to carry out the work of the Atlas program. By 1958, the work force had increased to 9,000, and by 1962, it was up to 32,500. General Dynamics made Astronautics a full division in 1961. Astronautics managed this rapidly expanding organization as a single project throughout the period.

However, with the development of different versions of the *Atlas,* and the development of new projects such as the *Centaur* upper stage and the *Azusa* tracking system, "priority problems were created in functional line departments, with resultant conflicts over authority and the jeopardizing of performance, scheduling, and cost." The Astronautics Division responded to this problem by "utilizing a program control plan called the 'matrix' system which provided a director for each program undertaken by the company." Program directors and department managers resolved priority conflicts.[84]

By 1963, Astronautics organized every new major program with the project system using the matrix structure. Atlas Program Director Charles Ames described the organization in the following way: "Under this system, the program director . . . is responsible for the successful accomplishment of the project . . . Generally, personnel working full-time on a project are assigned to the project line organization. The project line activities are organized to fit the specific task . . . Personnel not assigned to the project line organizations work in functional or 'institutional' departments. Institutional engineering maintains strong scientific and applied research groups as well as preliminary design and systems analysis groups."[85]

Required by the military, and prompted by their own complex projects, to institute project management, the contractors fit the weapon system concept and its project management into their organizations through the creation of matrix management. Matrix management provided companies with the means to move engineers across projects while maintaining disciplinary expertise, becoming the industry standard by the 1960s. Matrix management allowed the government to manage each project with systems management while also permitting each corporation to coordinate and control the multiplicity of projects in a way that was consistent with the corporation's overall strategy.[86]

Conclusion

The air force policy of concurrency kept the ballistic missile program on the fast track, in keeping with the perceived urgency of the Soviet threat in the 1950s. Industry reacted by developing matrix management, allowing several projects to be managed simultaneously. Unfortunately, in the desire to eliminate red tape and bureaucracy, Schriever's organization also removed many of the checks necessary to coordinate technical details and budget for large systems. The result was a series of missile failures, compounded by huge cost overruns.

To remedy this situation, the WDD's successor, the BMD—along with R-W and its successor, TRW—developed methods to improve missile reliability. These included exhaustive testing, component inspection and tracking, and configuration control to ensure that the design matched the hardware that was launched. Schriever's group expanded the methods of configuration control into configuration management, an important process that tied cost estimation to engineering changes. The air force modified contracts based upon the accepted changes.

By the early 1960s, the air force integrated these concepts into AFSC and its administrative processes for weapon system development and procurement. These processes incorporated systems engineering, systems analysis, reliability, and configuration control. Schriever's authority grew, soon encompassing the air force's space programs, but at the same time he gave up his independence from the air force bureaucracy. AFSC made systems management the standard for large-scale systems.

Schriever soon found that McNamara's centralizing changes, particularly phased planning, eroded his independence and slowed the pace of development. With one hand he fought the changes, and with the other he rolled with the punches and promoted managerial innovation within AFSC. Just as Schriever and his technical officers promoted standardization to better control their organizations, so too did McNamara, promoting standardization across the entire DOD. Schriever's methods, developed to centralize management of large-scale projects, provided the basis upon which McNamara could then control all military R&D programs.

The problems and solutions faced by the air force were not unique. During the 1950s, the army's JPL was a leader in missile development, confronting similar problems and developing similar solutions to its air force and navy competitors. JPL provides a good point of comparison with the air force experience, as it transitioned from military missiles to civilian satellites, and from Army Ordnance to NASA.

FOUR

JPL's Journey from Missiles to Space

> Pride in accomplishment is not a self-sufficient safeguard when undertaking large scale projects of international significance.
> —Kelley Board, after *Ranger 5* failure

The Jet Propulsion Laboratory (JPL), located in Pasadena, California, and managed by the California Institute of Technology, began as a graduate student rocket project in the late 1930s and developed into the world's leading institution for planetary space flight. Between 1949 and 1960, JPL transformed itself twice: first, from a small research organization to a large engineering development institution, and second, from an organization devoted to military rocketry to one focusing on scientific spacecraft.[1]

JPL's academic researchers did not initially recognize the many differences between a hand-crafted research vehicle and a mass-produced, easily operated weapon, or highly reliable planetary probes. The switch from research to development required strict attention to thousands of details. Properly building and integrating thousands of components was not an academic problem but an organizational issue. JPL's engineering researchers learned to become design engineers, and in so doing some of them became systems engineers.

Learning systems engineering on tactical ballistic missiles, JPL managers and engineers modified missile practices to design and operate spacecraft. The most significant missile practices that carried over to spacecraft were organizational: component testing and reliability as well as procedures for change control. A few JPL managers learned these lessons quickly. However, it took a number of embarrassing failures for JPL's academically oriented engineers and managers to accept the structured methods of systems management.

JPL independently recreated processes that the air force developed on its ballistic missile programs: systems engineering, project management, and configuration control. The history of the two organizations shows that the processes were the result of not individual idiosyncrasies but larger technical and social forces.

From Student Rocketry to Weapons Research

In 1936, Caltech graduate student Frank Malina learned of Austrian engineer Eugen Sänger's proposed rocket plane. This stimulated Malina's interest in rocketry, and aeronautics professor Theodore von Kármán agreed to serve as his thesis adviser. Learning little from a visit to secretive rocket pioneer Robert Goddard, Malina and his assistants began rocket motor tests in an isolated area near Pasadena. After several failures, they succeeded in running a forty-four-second test firing. In May 1938, a new heat-resistant design operated for more than a minute.[2]

In 1938, Army Air Corps commander H. H. "Hap" Arnold made a surprise visit to Caltech and took interest in the project. He asked the National Academy of Sciences to fund research on rocket-assisted aircraft takeoff. Von Kármán got the job, with Malina doing most of the work. Money soon began to flow, and by July 1940 the group moved permanently from the Caltech campus to the test site.[3]

Malina's growing team used theoretical analysis and practical experimentation to create a series of technical breakthroughs that became the foundation of solid-propellant rocketry. The army and navy wanted rockets to assist aircraft takeoff from short airfields and aircraft carriers, leading Malina's team to consider mass production of the rockets. Malina unsuccessfully tried to interest aircraft companies. Failing in this, he, von Kármán, and others started the Aerojet Company, which by 1943 had large navy orders. Although JPL developed the initial designs, it never had to deal with manufacturing problems, passing these to Aerojet.[4]

After the military discovered German preparations to launch *V-2* rockets, some army officers paid greater attention to rocketry. The Army Air Forces was not interested because long-range rockets did not promise an immediate payoff and because it had a vested interest in manned bombers. By contrast, Army Ordnance officers saw rockets as long-range artillery and hence as a means to extend the range of their artillery and political aspirations. They urged Caltech to propose a comprehensive program, which led to the official founding of JPL in June 1944 with an Army Ordnance contract for $1.6 million. Despite Caltech leaders' initial view that JPL would aid the army during only wartime, they quickly became addicted to the contract's overhead money. JPL became a permanent operation.[5]

JPL proposed to build a series of progressively larger and more sophisticated rockets, named in rank order *Private, Corporal, Sergeant,* and *Colonel. Private,* developed in 1944 and early 1945, proved successful when designed as a simple rocket but inaccurate when modified to include wings. *Private*'s performance proved not only that JPL could design a simple rocket without attitude control or guidance but also that long-range rockets were impractical until JPL developed an automatic control system. The Corporal series began with an unguided sounding rocket known as the *WAC Corporal,* intended to achieve the highest possible altitude. It reached altitudes of forty miles in October 1945 and was the immediate progenitor of Aerojet's *Aerobee* sounding rocket, used for years after as a scientific research vehicle. Relations between JPL and Aerojet were good, as JPL researchers passed research innovations to Aerojet, which developed them for production. With financial interests in Aerojet, JPL researchers benefited handily.[6]

The organization of JPL's early rocketry was simple. It began as a student research project, with Malina, John Parsons, and Edward Forman constructing test stands, motors, and fuels. The group added a Research Analysis section, which performed parametric analyses of aircraft takeoff with rocket assistance and developed design objectives. Homer Stewart and Hsue-shen Tsien did many of these tasks, which Stewart later recalled as being the systems engineering for the group.[7]

As the program grew, Malina directed JPL while Army Ordnance handled the coordination among JPL, White Sands Missile Range, the Signal Corps, and the Ballistic Research Laboratory of Aberdeen Proving Ground. The latter two organizations assisted with flight test data acquisition. Malina divided JPL's twenty-two personnel into seven small groups: Booster, Missile, Launcher and Nose, Missile Firing, External Ballistics, Photo and Material, and Transportation and Labor. The army's contingent totaled thirteen. Prior to each test round, Malina held a conference where each group discussed prior results and checked weather and preparations. Douglas Aircraft manufactured the rocket, but the team often performed last-minute modifications at White Sands.[8]

WAC Corporal paved the way for JPL's first true surface-to-surface missile, the larger and more complex liquid-fueled *Corporal E.* JPL engineers developed a comprehensive test program to ensure that the components and the integrated vehicle functioned correctly. They developed static structural tests, hydraulic tests for all fluid flow components, and rocket motor tests. Engineers also created a full-scale model used to check pressure and temperature characteristics under firing conditions on a static test stand at Muroc, California. The test stand held the vehicle on the ground as the engine fired, while electrical instrumentation measured structural loads, pressures, and temperatures. Douglas Aircraft manufactured the flight test vehicles, which the army transported to its new assembly and launch facilities at White Sands, where engineers performed final leak and electrical tests. Technicians then moved the rocket seven miles to the launch site, where the crew simulated a firing for training purposes and as a final telemetry check. They then fueled and launched the vehicle.[9]

JPL engineers fired the first *Corporal E* in May 1947. The first round was a success, but round two produced insufficient thrust. Round three failed when the rocket motor throat burned out and the control system failed. Engineers went back to the drawing boards. Only in June 1949 did the next *Corporal E* fly, with a new design using axial-flow motors.[10]

After the Soviets exploded their first atomic bomb in August 1949, Army Ordnance officers asked JPL Director Louis Dunn[11] and Electronics Department head William Pickering whether *Corporal* could be converted into an operational missile. Dunn stated that JPL could handle this conversion if it developed a guidance and control system from existing technologies. In March 1950, Army Ordnance decided to make *Corporal* into a weapon.

When the Korean War broke out in the summer of 1950, the Truman administration gave Chrysler executive K. T. Keller the charter to develop missiles as quickly as possible. Rejecting a Manhattan Project–style program, Keller decided instead to exploit existing missile programs that held promise. *Corporal* was the army's best-developed missile, so Army Ordnance committed it to rapid development. With this decision, JPL embarked upon a venture that changed it from a research institution into the equivalent of an army arsenal.[12]

From Weapons Research to Weapons Development

Corporal's acceleration shifted JPL's emphasis from engineering research to large-scale development. Fundamental research continued, but at only a fraction of JPL's budget, as engineering and production budgets for the Corporal project climbed dramatically.

Dunn and Pickering led the Corporal project. Dunn came to Pasadena in 1926 from South Africa to study aeronautical engineering under von Kármán, completing his doctorate in six years. Associates characterized him as a decisive, orderly executive who organized JPL on a project-oriented model. Pickering came to Caltech from New Zealand, taking a Ph.D. in physics in 1936, then joining the electrical engineering faculty. He worked on electronics for cosmic ray studies and expanded into telemetry and guidance design in 1944 at JPL. In 1950, Pickering was head of JPL's Electronics Department, along with being the Corporal project's director. Pickering preferred an academic orientation, emphasizing technical excellence organized through working groups.[13]

In October 1951, JPL froze *Corporal*'s aerodynamic configuration, and Army Ordnance committed *Corporal* to limited production, selecting Firestone Tire and Rubber Company to manufacture the missile. JPL expanded its staff and used the production missiles for test firings.[14] Test firings were disappointing, as electronic components frequently failed. Although JPL engineers realized that rocket engine vibrations were a factor, they had seriously underestimated the magnitude of the problem. As failures mounted, they began recording failure statistics. By summer 1953, with more than forty firings completed, JPL calculated missile reliability at only 48%, even with the electronics wired into two strings of components such that if a failure occurred in one set, the other would continue to operate.[15] Reliability problems threatened to undermine the project and with it the credibility of Army Ordnance officers and JPL engineers.

The majority of early *Corporal* failures were in electrical systems such as power, guidance, radar, and telemetry. According to JPL engineers, "Neither the reason for the failure nor the specific part failing is known in most instances; failures are commonly attributed to vibration." Consequently, they tested components with a sine-wave vibration generator and at high and low temperatures. They found that some components had resonant frequencies that could lead to physical breakage. Engineers developed a program to test all electronic parts and changed both vendors and parts. They developed a rule of thumb to repair failures on the spot but redesign a component if it failed three times.[16]

Whereas aircraft structural vibrations typically occurred at predictable frequencies matching the rotation rate of propellers or jet engine rotors, rocket engines produced vibrations at near-random frequencies. In addition, high speeds and changing altitudes placed strong, highly variable aerodynamic forces and vibrations on the missile's structure. These shook loose or severed wires, connectors, and soldered components, causing electrical short circuits and intermittent connections. The failures raised havoc with the electrical systems such as radio guidance, attitude control, and telemetry subsystems.[17]

One response was to acquire better flight data. Engineers placed accelerometers and strain gauges on the missile, and they sent the data through the radio system to be recorded on the ground. Because the speed of data collection and radio transmission was too slow to capture the full profile of high-frequency vibrations, engineers constructed algorithms to calculate vibration frequencies and amplitudes from the infrequent data samples. These algorithms were sufficiently complex and data-intensive to require the use of an IBM programmable computer. After much work on these data transmission, storage, and processing problems, engineers found vibrations to be highly unpredictable.[18]

Because of the expense and inconclusive flight test results, engineers constructed a vibration simulator to test individual components and component packages. They expanded component and package testing, and they formulated guidelines and standards.[19] Vibration testing henceforth became a standard element of component qualification and missile development.

JPL engineers also theoretically analyzed missile reliability. Assuming that each component had a measurable failure rate, engineers estimated missile reliability simply by multiplying component reliability estimates. For example, for a missile with only two electronic components, where these components both had to operate and each had a 90% probability for successful operation, multiplying their reliability estimates together gave a combined reliability estimate of $0.9 \times 0.9 = .81$, or 81%. In this way, engineers estimated the decrease in reliability as they added electrical components. With calculations such as these, engineers determined that adding a second parallel "string" of components significantly improved missile reliability.[20]

In light of the Korean War and the tense situation in Europe, the army decided to deploy *Corporal* despite its severe reliability problems. Both JPL and the army soon realized that *Corporal* was not designed for operations. JPL engineers had initially designed *Corporal* purely as a research vehicle using World War II vintage hardware, much of it out of production. When failures occurred, researchers investigated and fixed them on the spot. The army sent military crews to White Sands to learn how to prepare and fire the missiles, but they did not have the expertise of professional engineers. JPL's lack of operations experience showed in its poor documentation and frequent design changes. Poor training led to more failures and lower reliability, because operational reliability depended upon enlisted personnel to maintain and fire the missile.

Pushing *Corporal* into crash production aggravated the situation because the army had to use JPL's sensitive laboratory equipment in the field. Many missiles failed tests because ground equipment had shifted out of tolerance. Even after relaxing tolerances, experience showed that on average, in the four hours necessary to prepare and fire a missile, one electronic component failed. The awkward, bulky equipment was extremely cumbersome. When a *Corporal* battalion moved, its convoy stretched sixteen miles![21]

Flight instrumentation was another major problem. JPL engineers initially believed they needed little instrumentation for the tactical missile. This was a mistake. Jack James, an engineer assigned to this problem, reported that throughout a program of 111 development firings and 150 training rounds through June 1957, engineers and technicians modified every missile to install instrumentation. With thirty tactical ground systems capable of firing

Corporal but only eleven telemetering stations and two telemetry processing facilities (one at JPL and one at White Sands), telemetering stations had to be shipped to the firing site and all data sent to JPL or White Sands for analysis. Because JPL had few personnel trained to analyze test data, substantial delays ensued. These experiences taught James that good vehicle design required up-front consideration of testing and operational factors, with a design that incorporated sufficient instrumentation.[22]

As problems mounted, in November 1953 the army proposed to assign a commanding officer to JPL, a significant step toward turning it into a government arsenal. If the army wanted to control JPL, the trial balloon was ill-timed because it had little choice but to rely upon JPL to rapidly deploy *Corporal.* Although technical problems reduced JPL's credibility, the urgency of speedy deployment weakened the army's position even more. Army Ordnance backtracked and in 1954 gave JPL even more responsibility for *Corporal.* Because JPL was the only organization capable of making the missile work, army officers had little choice.[23]

The best efforts of JPL and the army improved *Corporal*'s reliability to an estimated 60%, the best achievable with its inherent design deficiencies. This left serious doubts as to its utility.[24] *Corporal*'s real value was that it trained the army and JPL how to, and more importantly, how *not* to develop a missile. From this experience, JPL's leaders recognized that academic, ad hoc design methods and loose organization were not sufficient to create an operational weapon. They vowed that on the next project, they would not repeat these mistakes.

Applying the Systems Approach

By mid-1953, JPL's continuing research in solid-propellant rocketry led to the conclusion that solid propellants could equal or exceed liquid propellants in performance as well as eliminate the cumbersome logistics of liquid-propelled missiles. Following up on this conclusion, Army Ordnance funded several studies, from which it selected JPL's *Sergeant.* JPL managers and engineers stressed their recent recognition that missiles had to be viewed "as true system problems" that considered ground-handling equipment, operations, and training as well as technical improvements such as an improved guidance system and solid-propellant propulsion. Warning Army Ordnance about the dire consequences of making Sergeant a crash program, JPL Director Louis Dunn stated that "a properly planned development program" would "pay for itself many times over" by avoiding changes to production and operations.

Shortly after the army accepted JPL's proposal, Dunn left to head Ramo-Wooldridge's Atlas project. Corporal project manager William Pickering became JPL's new director in August 1954. Pickering reorganized the laboratory to mirror academic disciplines on the Caltech campus.[25]

Even though he structured JPL on an academic model, Pickering recognized some of its limitations. Noting, "R&D engineers may not necessarily fully appreciate military field conditions," Pickering assigned "certain personnel a particular system responsibility as a sole task." They performed studies of training, logistics, organization, and other factors to determine the "instrumentation, training and schooling requirements, the caliber of personnel requirements, and a typical Table of Organization for the missile battalion."[26] Pickering assigned Robert Parks as project manager and Jack James as Parks's deputy. James soon developed processes that would significantly change JPL's management practices.

Jack James graduated from Southern Methodist University in 1942 and began his career at General Electric (GE) in Schenectady, New York. Starting by working on turbine engines, he soon transferred into the Test Engineering program, where he rotated through a number of laboratories and projects to gain experience. During World War II, he served as a navy radar officer on the battleship *South Dakota*. After the war, he returned to GE.

At GE, James worked for Richard Porter on the Hermes project to test-fire modified *V-2* rockets. At the end of World War II, Porter had worked on the Paperclip project, which brought German rocket engineers and technicians to the United States, and Porter brought a number of the Germans with him to GE. GE developed the radar guidance system, and James worked with *SCR-584* radar systems, on which he "had the chance to make many mistakes." In 1949, after the Research and Development Board picked JPL to manage the Corporal project, James moved to Pasadena. He had a "nightmare job" getting GE to deliver the guidance system, because GE had hoped to manage the project and had "lost heart in the job." After Dunn left JPL, James helped complete *Corporal*.[27]

One of *Corporal*'s irritants was its lack of instrumentation for telemetry data. James, who was the project manager for the first two *Sergeant* flights, designed instrumentation into the new missile for testing and troop training, even though this added extra weight. Engineers could reconfigure telemetry equipment and measurements, depending upon the missile's use for engineering development, testing, or training—or for its final military purpose.[28]

Another of *Corporal*'s faults was horrendous reliability and maintenance. *Sergeant* incorporated the vibration testing established on *Corporal* for components. James also investigated the maintenance problem theoretically, to determine the best design, procedures, and supply inventories. He noted that some branches of the army recommended that suppliers create test equipment to isolate faulty components down to the piece-part level. In contrast, his analysis showed that small numbers of larger replaceable packages were more cost-effective. Because the army levied stringent reliability requirements, *Sergeant* engineers developed a strict failure reporting system that required documentation about how engineers would permanently repair each failure.[29]

To Pickering, Parks, and James, the systems approach meant including reliability, testing, and maintenance early in the design process. Sperry Rand Corporation, which the army selected to manufacture *Sergeant,* created a systems engineering program for test equipment. It consisted of formal and informal meetings and conferences, coordination of engineering changes, and the development of consistent testing, reliability, and maintenance methods at JPL and Sperry and in the army. *Sergeant* managers and engineers standardized environmental testing standards, safety procedures, component mounting practices, and maintenance procedures. They also separated testing into five major phases: feasibility flights, guidance system development, system development and integration, engineering model flights, and system proof tests.[30]

JPL used old and developed new organizational structures and procedures in its relationship with Sperry. Army Ordnance defined institutional arrangements, using JPL as the contractor responsible for technical research, development, and cognizance. Sperry was to manufacture the missile as the prime contractor, but not until it learned how to build the system as co-contractor with JPL. JPL engineers issued Technical Guidance Directions, and Sperry next provided cost estimates. With JPL's approval, Army Ordnance officers then funded Sperry on a cost-plus-fixed-fee basis. The army required two reviews, a Design Release Inspection and a Design Release Review, both held early in 1959.[31]

Because of the planned transition from JPL to Sperry, James required that JPL engineers describe their designs in a series of documents that James sent to Sperry. This forced JPL engineers to synchronize design work to a fixed schedule and to produce consistent documentation. If an engineer was unsure about how a design interacted or connected to a neighboring subsystem, that engineer would simply check the design document's latest release. James also instituted a system of document change control so that engineers could not arbitrarily change their designs. Modifications would pass through James, who would ensure design and documentation consistency through a Research Change Order.[32] This progressive design freeze, augmented with change control, turned out to be one of the most significant organizational elements in the success of *Sergeant.*

Engineering changes were a prominent source of conflict between JPL and Sperry. Coordination between the two started in 1956, with Sperry assigning a number of engineers to work with JPL in Pasadena. In 1957, monthly coordination meetings that alternated between Pasadena and Sperry's new Utah facility began. After negotiations with Sperry, JPL managers extended their Research Change Order system so that it governed engineering and production changes at JPL and Sperry. That same year, the two organizations created a biweekly Operational Scheduling Committee that initially governed the scheduling and preparations of test rounds but soon included broader coordination and contractual issues. Continuing problems led to a project-based reorganization at Sperry, and both organizations established Resident Offices at each other's facilities. The Sergeant Action Review Committee, formed in December 1959, reviewed all design changes, allowing only those that were mandatory.[33]

On Sergeant, JPL proved its capability as an army arsenal, with full capability to design, develop, and oversee a missile from inception to operational deployment. JPL engineers developed the procedural expertise necessary to convert research technology into operational weapons, including reliability and maintenance, systems analysis, project scheduling and coordination, and phased planning. JPL Director William Pickering supported these systems methods, although he clung to an academic-style organization. Contractual relationships between the army, JPL, and Sperry led to the development of formal systems to report and respond to failures, and to progressively freeze and document the engineering design as it progressed. Jack James recognized their utility to coordinate diverse design activities and would apply them again on spacecraft projects, as JPL underwent its second major transformation from an army arsenal to a National Aeronautics and Space Administration (NASA) field center.

From Missiles to Space

JPL's entry into the space program came through an alliance with the Army Ballistic Missile Agency (ABMA) on the Jupiter intermediate-range ballistic missile program. In 1955 and 1956, JPL worked with ABMA on a backup radio guidance system and the reentry test vehicle for *Jupiter.* The radio guidance work gave JPL the funding and opportunity to improve radio communications between ground systems and flight vehicles, later evolving into JPL's Deep Space Network. The reentry test vehicle was a spacecraft in all but name. ABMA and JPL performed reentry test flights between September 1956 and August 1957. The Army Ordnance commander, Gen. John Medaris, ordered the remaining rocket hardware to be put into storage, hoping to launch a spacecraft.[34] For the moment, Medaris had to wait; the navy's *Vanguard* was to launch the first U.S. spacecraft. However, the failure of *Vanguard*'s first test flight in December 1957 paved the way for the army.[35]

With public pressure building in the wake of *Sputnik,* President Eisenhower gave the army the green light to unleash Wernher von Braun's ABMA team and Pickering's engineers at JPL. Pickering seized the opportunity. By participating in the space race, Pickering could return JPL to engineering research instead of the drudgery of weapon systems development. In a brief discussion immediately preceding a meeting to assign responsibilities for the orbital attempt, Pickering convinced Medaris to assign JPL the spacecraft and tracking network. JPL engineers quickly designed a high-speed stage, eventually designated *Explorer 1,* consisting of clusters of *Sergeant* solid motors and a cylindrical can that contained telemetry equipment and scientific experiments.[36]

JPL engineers used processes developed on *Sergeant* and the reentry test vehicle. By the summer of 1956, ABMA and JPL had tested rocket motors in small vacuum chambers to ensure that they would operate in space. Engineers expanded these tests to examine the *Explorer* spacecraft's capacity to withstand large temperature variations in a vacuum, such as it would encounter when in the Sun or in the shade of the Earth. JPL engineers replaced vacuum tubes with transistors, repackaged electronic components, and tested the entire package with random-vibration tests. In addition, they used redundancy to increase the chances for success if one component failed. JPL's ground telemetry systems were ready. When in January 1958 the ABMA's *Jupiter* rose from Cape Canaveral, JPL's *Explorer 1* spacecraft and ground systems functioned perfectly, returning scientific data leading to the discovery of the Van Allen radiation belts and confirming that micrometeorites were not a problem.[37]

ABMA and JPL followed *Explorer 1* with a series of spacecraft in the Explorer and Pioneer series. Because the primary goal was to compete in a prestige race with the Soviets, engineers hurriedly lashed together existing technologies to jury-rig space missions. *Explorer 2–Explorer 6*, *Pioneer 3*, and *Pioneer 4* had a mixed record, with several successes and several failures. Because of the urgency of the space race, neither the army nor Congress questioned this record. Spacecraft failures occurred out of sight, unlike spectacular rocket explosions and their unpleasant publicity. JPL engineers, used to the army mentality of firing many test rounds, thought of these early spacecraft as test rounds and were not overly concerned with achieving a perfect record. They rushed into space and reverted to the earlier Corporal mentality of small project groups using informal methods.[38]

Despite the exploits of ABMA and JPL, the army lost its battle against the air force and the new NASA for a significant space role. On January 1, 1959, President Eisenhower transferred JPL to NASA, and the ABMA soon thereafter.[39]

For NASA, JPL proposed a new program for lunar and planetary exploration known as Vega. Vega was to develop a third-stage rocket and spacecraft similar to *Explorer*'s high-speed stage and payload. Its spacecraft design was far more complex than *Explorer*'s because it needed to operate for months in transit to the Moon, Venus, or Mars. The *Vega* spacecraft was to feature important new technologies, including solar panels, three-axis attitude stabilization, and a flight computer. Just as after *Corporal,* when JPL managers and engineers planned the *Sergeant* missile as a "systems job," JPL engineers and managers carefully planned for *Vega,* succeeding the hastily built *Explorer* and *Pioneer* spacecraft.[40]

JPL Director Pickering selected Clifford Cummings as Vega project director. Cummings had worked under Pickering on Corporal and Sergeant, developing analytic tools. He believed that better maintenance required better analysis of training programs and costs, supply networks and logistics, test equipment, and vehicle design. Vocal and outspoken, Cummings believed that scientists and engineers could work out difficult problems through working groups and a thorough test program.[41]

Cummings and his deputy, James Burke, organized Vega's test program using lessons from Corporal, Sergeant, and Explorer. He and Burke planned a mockup spacecraft for structural and mechanical tests as well as an engineering model for environmental and electrical tests. Only after the engineering model passed these tests would JPL build the flight spacecraft. *Vega* featured a new "systems test" that would simulate the flight sequence and events with all of the spacecraft subsystems working together. Engineers were to record test results on specialized forms for later analysis. After engineers assembled and tested the spacecraft in this manner, they would then perform the same tests in a large vacuum chamber, then in a vibration test facility, and finally at Cape Canaveral prior to launch.[42]

Plans for *Vega* did not come to fruition because NASA Administrator Keith Glennan canceled the program in December 1959 to avoid duplication of the air force's previously secret *Agena* upper stage. Glennan decided to use the air force's *Atlas-Agena* for NASA's early missions instead of *Vega.* Never again would JPL work on the rocket designs upon which it had made its reputation. In place of Vega, JPL acquired NASA's robotic lunar and planetary missions, which became the Ranger, Surveyor, and Mariner programs. Time spent planning for *Vega* was not completely wasted, as its design studies and test plans carried over to *Ranger.*[43]

Functional Management or Project Management?

JPL's lunar and planetary programs developed under very different organizational regimes. In the lunar program, Cliff Cummings and James Burke ran the Ranger program on an academic model; Burke coordinated the activities of the subsystem engineers who worked under the technical division chiefs. JPL contracted with Hughes Aircraft Company (HAC) to design and build the *Surveyor* lunar lander. *Surveyor* lacked support from JPL, whose personnel concentrated on *Ranger* and *Mariner*. Because of JPL's neglect, HAC ran the program as it saw fit. By contrast, Robert Parks and Jack James ran the planetary program, which consisted initially of the *Mariner* spacecraft to fly by Venus, on the formal model they had developed on *Sergeant*. Although *Mariner's* design was a modification of *Ranger*, the spacecraft achieved quite different results: disastrous failures on *Ranger* and spectacular success on *Mariner*. Their contrasting fates illustrate the significant influence of organizational structure and processes on the technical success or failure of spacecraft.

Ranger and *Surveyor* were intended to support NASA's lunar program both by attaining space achievements before the Soviets did and by helping the Apollo mission. *Ranger* was to take close-up pictures of the lunar surface before the spacecraft crashed onto the Moon and to help engineers develop spacecraft technologies for use on other programs. *Ranger* had an additional goal: "to seize the initiative in space exploration from the Soviets." *Surveyor* was to perform a soft landing on the lunar surface. Conflict between scientific and engineering goals hampered both projects. Scientists desired on-board experiments, but for engineering purposes and for Apollo support, experiments were a nuisance. By contrast, Mariner was a purely scientific program to explore Venus and Mars, relaying photographs and scientific data back to Earth. It did not support Apollo and, consequently, had clearer mission objectives.[44]

By late 1959, some of JPL's managers believed that JPL needed to change its organizational structure. They thought that Pickering's academic structure did not work well for large projects. To investigate, JPL hired management consulting firm McKinsey and Company to assess JPL's organization. Based on the firm's recommendations and pressure from managers like Jack James, Pickering established project-oriented Lunar and Planetary Program offices, but he maintained the authority of JPL's functional divisions and added the Systems Division. Pickering selected Cliff Cummings to head the Lunar Program Office. Cummings, in turn, selected his protégé James Burke to head Ranger.[45]

Burke, a Caltech mechanical engineer, had a reputation as a brilliant engineering researcher and technical specialist. He had an easygoing attitude with others but drove himself very hard.[46] On Ranger, which began in December 1959, Burke's mild demeanor turned out to be a handicap. The 1959 reorganization created project managers, but the division chiefs from Pickering's functional organization controlled the personnel. Project managers had little authority and had to negotiate with powerful division chiefs for personnel and support. Burke did not have the authority to force division chiefs to abide by project decisions. For example, when Mariner needed personnel, division chiefs compromised Ranger by transferring some of Ranger's most experienced engineers to the more glamorous Mariner. Biweekly meetings with the divisions focused on program status and scheduling, not technical problems or systems engineering.[47] Burke's project office consisted of a single deputy, and he gave the critical systems engineering tasks to the Systems Division, ad hoc committees, and technical panels.

Although JPL had developed substantial expertise in reliability on Corporal and Sergeant, reliability and quality assurance engineers could only advise design engineers, who could reject their advice. With many senior engineers transferred to Mariner, Ranger's reliability suffered. Design inconsistency resulted from continuing changes in the scientific experiments requested by NASA headquarters. Ranger also suffered from a requirement to sterilize components by baking them at high temperatures, which significantly reduced electronic component reliability.[48]

Burke's lack of authority inside JPL was a small problem compared to his lack of authority over external organizations. JPL reported to the Office of Space Flight Programs at NASA headquarters. Air force *Atlas* and *Agena* vehicles were to launch *Ranger*, yet launch vehicles fell under the jurisdiction of the Office of Launch Vehicles at NASA headquarters, which assigned responsibility for *Atlas* and *Agena* to Marshall Space Flight Center (MSFC). MSFC had responsibility for, but no authority over, the air force for these vehicles. Thus, authority for the Ranger program was divided between two NASA field centers, two headquarters offices, and NASA and the air force.[49]

Ranger was not a priority for MSFC or for the air force. MSFC was busy designing the *Saturn I* rocket, a step toward von Braun's dream of manned space flight. For the air force, NASA's use of *Atlas* and *Agena* was secondary to developing ballistic missiles. *Agena* contractor Lockheed gave priority to the hundreds of upper stages slated for the air force as opposed to the nine purchased by NASA. NASA did not help matters, assigning responsibility for the Agena program to a headquarters-chaired committee, where *Ranger* was only one of several NASA *Agena* users. Project personnel had to work through the committee, which reported to MSFC, which in turn coordinated with the air force, which then directed Lockheed.

With this confusing organization, launch vehicle problems were a virtual certainty. Having many organizations interposed between JPL and Lockheed led to misunderstandings about the electrical and physical connections between JPL's spacecraft and Lockheed's *Agena* upper stage. Exasperated JPL engineers could not get crucial *Agena* information from the air force or Lockheed because they did not have the "need to know" required by air force security.

In September 1960 Lockheed sent a mockup of the *Agena* upper-stage interface hardware to JPL. Not surprisingly, the hardware did not match JPL's expectations. After the ensuing investigation, the air force granted security clearances, then let NASA sign its own contract with Lockheed in February 1961. The problems also led managers and engineers in the air force and NASA to hold a design review covering interface hardware in December 1960. JPL engineers sent tooling and spacecraft mockups to Lockheed to check interface designs, so that when they manufactured flight spacecraft, they would match *Agena*'s interfaces.[50]

Mariner was JPL's showcase project, intended to fly two spacecraft past Venus in 1962 and two more past Mars in 1964. Originally slated to launch with the new high-energy *Centaur* upper stage, NASA canceled the *Mariner A* spacecraft (the first of the series) when it became clear that *Centaur* would not be available in time. Regrouping, JPL engineers lightened the design to launch on *Atlas-Agena* launch vehicles. NASA approved the new *Mariner R* spacecraft in the fall of 1961. Mariner benefited from its allure as a planetary mission and from its stable complement of onboard science experiments.[51]

Although Mariner's organization included elements similar to Ranger's, a number of features significantly strengthened Mariner's management. As with Ranger, project managers emphasized interfaces between the spacecraft and launch vehicle, required significant testing, used JPL's matrix structure, and had a small project office. There the similarities ended. Robert Parks, former Sergeant program manager, ran JPL's planetary programs, and he selected his Sergeant deputy, Jack James, as project manager. The two Sergeant veterans decided to use Sergeant's best management features, particularly failure reporting, design freezes, and change control. Mariner engineers began by writing functional specifications to resolve spacecraft interface problems. They then created a design specification manual that defined the preliminary design, mission objectives, and design criteria. James created a development operations plan outlining the communication processes for the project, including interfaces, technical design decisions, schedule reporting, and design status meetings. James's plan even specified what topics each status meeting should cover and who should attend. Unlike on Ranger, on Mariner James tracked the development of specifications and design drawings, not just hardware.[52]

James believed that the most innovative management feature of Mariner was the use of progressive design freezes. After a survey of subsystems to determine when to freeze each design element, the project periodically published a *Mariner R* change freeze document, along with any approved changes to drawings or specifications. Once frozen, a component's design could be modified only through an engineering change requirement form approved by James.

Problem reporting became one of the project's significant innovations. Project manager James instituted the "P list," a list of critical problems. Any problem that made the P list received immediate attention and extra resources. The project implemented a failure reporting system for *Mariner* in November 1961, starting with system integration tests for the entire spacecraft. Failure reports were distributed to division chiefs, the project office, and engineers responsible for designing components and subsystems.[53]

As JPL prepared for its first *Ranger* and *Mariner* flights, its engineers and

Mariner Venus 1962, also called *Mariner 2.* Mariner's success helped convince NASA to reform JPL rather than reject it. Courtesy NASA.

managers were confident that they would succeed. Even if faults occurred, five *Ranger* test flights and two *Mariner* spacecraft gave ample margin for the unexpected. Despite last-minute changes to Ranger's science experiments and occasional testing glitches, both projects remained on schedule. The Ranger project planned to build five "Block 1" spacecraft, only one of which had to work properly for Ranger's initial objectives to be met. Surely one of them would.

The NASA School of Hard Knocks

In the summer of 1961, JPL engineers prepared the first *Ranger* spacecraft for flight. Having successfully passed through a series of structural, electrical, and environmental tests, *Ranger 1* was scheduled for launch in July. After two hardware component failures in the spacecraft and one in the launch vehicle delayed the launch, the *Atlas-Agena* took off in August. Within minutes, engineers found that the *Agena* upper stage did not ignite, stranding *Ranger 1* in Earth orbit. The spacecraft operated properly but burned up in Earth's atmosphere on August 30. *Ranger 2* launched in November. Its *Agena* also failed, placing *Ranger 2* in a low orbit, from which it soon disintegrated in the Earth's atmosphere. The air force launched an investigation of the failures and pressured Lockheed for solutions.[54]

Ranger 3 was the first of JPL's Block 2 design, which included new science experiments as well as a television camera to take lunar surface photographs for Apollo. In January 1962, the air force's launch vehicles placed *Ranger 3* on a trajectory that would miss the Moon, but JPL decided to operate the spacecraft as close as possible to the normal mission. While performing the first-ever trajectory correction maneuver (firing thrusters to change the spacecraft's course), a reversed sign between the ground and flight software resulted in the spacecraft's course changing in exactly the opposite direction from that desired. Two days after launch, the spacecraft computer failed, ending the mission prematurely.[55]

Launching in April 1962, *Ranger 4*'s computer failed almost immediately after separating from *Agena*. This made the spacecraft blind and dumb—able to send no data to Earth and unresponsive to commands. Its trajectory was nearly perfect; the spacecraft crashed onto the Moon's surface three days later. Although NASA and the newspapers proclaimed the mission a great success, which it was for the air force's launcher, it was a complete technical failure for JPL. Engineers hypothesized that the failure was due to a metal flake floating in zero gravity that simultaneously touched two adjacent electrical leads.[56]

JPL next scheduled its two *Mariner R* spacecraft for launch. The first launch on July 22 failed after five minutes, when *Atlas*'s guidance system malfunctioned. Air force and General Dynamics engineers traced the problem to a missing hyphen in a line of software code. *Mariner 2* launched in August and successfully flew past Venus in December. Its outstanding performance kept Congress and NASA headquarters from losing all confidence in JPL.[57]

Burke and his team made several changes to *Ranger*'s design and organization between *Ranger 1* and *Ranger 5*. On early flights, division chiefs interfered with spacecraft operators, so Burke reserved authority to command the spacecraft to project managers and engineers. Engineers also added redundant features to the spacecraft. Despite proof that heat sterilization compro-

Ranger. The failures of the first six Ranger flights led to NASA and congressional pressure to strengthen project management. Courtesy NASA.

mised electrical component reliability, lunar program chief Cliff Cummings maintained NASA's sterilization policy. In any case, *Ranger 5*'s components had already been subjected to sterilization. Project engineers formed investigation groups to study interfaces and spacecraft subsystem interactions. They also added a system test to check interfaces and interactions between the spacecraft and flight operations equipment and procedures.[58]

These changes did not completely resolve the project's problems. *Ranger 5* launched in October 1962 and began to malfunction within two hours. This time, the power system failed, losing power from the solar panels. Shortly thereafter the computer malfunctioned, and the battery drained within eight hours, causing complete mission loss. The spacecraft's disastrous failure convinced JPL and NASA headquarters managers that the project was in serious trouble. Both launched investigations.[59]

The headquarters investigation board's findings left no doubt that JPL's organization and management of Ranger were at fault. They criticized JPL's dual status as a contractor and a NASA field center, noting that JPL received little or no NASA supervision. Project manager Burke had little authority over JPL division chiefs or the launch vehicles and no systems engineering staff and showed little evidence of planning or processes for systems engineering. JPL's approach—using multiple flight tests to attain flight experience—had strong traces of its army missile background. The board believed this "multiple shot" approach was inappropriate. Spacecraft had to work the first time, and testing had to occur on the ground, not in flight. The board characterized Ranger's approach as "shoot and hope." Supervisors left design engineers unsupervised, and design engineers did not have to follow quality assurance or reliability recommendations.[60] The Board's position was clear: "A loose anarchistic approach to project management is extant with great emphasis on independent responsibilities and individual accomplishment . . . This independent engineering approach has become increasingly ingrown, without adequate checks and balances on individual actions. Pride in accomplishment is not a self-sufficient safeguard when undertaking large scale projects of international significance such as JPL is now undertaking for NASA."[61]

JPL's rapid growth also caused problems. From 1959 to 1962, JPL's budget grew from $40 million to $220 million, while its staff grew from 2,600 to 3,800. Most of the new budget went into subcontracts, yet JPL's business management capability did not expand accordingly. Often JPL personnel "failed to penetrate into the business and technical phases of subcontract execution." This was due in part to JPL's "two-headed" nature, as a contractor to the government, and as a de facto NASA field center. NASA did not require JPL to use NASA or Department of Defense regulations, treating it as a contractor. On the other hand, NASA headquarters did not supervise JPL tightly like a contractor, because of its quasi–field center status.[62] Ranger managers had neither the authority and resources to carry out their mission nor the oversight that might lead to discovery and resolution of problems.

Failure in the race for international prestige was no longer acceptable to NASA headquarters.[63] The board recommended strengthening project management, establishing formal design reviews, eliminating heat sterilization, assigning launch vehicles either to NASA or to the air force, instituting a failure reporting system, clarifying Ranger's objectives, and eliminating extraneous features. To spur JPL into action, the board recommended that NASA withhold further projects until it resolved Ranger's problems.[64]

JPL Director Pickering quickly replaced lunar program chief Cummings with Robert Parks and project manager Burke with former Systems Division chief Harris Schurmeier. He gave them the authority that Cummings and Burke only dreamed of: power over division chiefs in personnel matters. He also created an independent Reliability and Quality Assurance Office with more than 150 people and gave this office authority over the engineers.[65]

Schurmeier instituted process changes to strengthen Ranger's systems engineering. He levied an immediate design review, adopted Mariner's failure reporting system, and ensured that engineers took corrective action. Schurmeier also instituted Mariner's system of engineering change control and design freezes. Ranger expanded the use of its Design Evaluation Vehicle to test component failure modes and inter-subsystem "cross-talk and noise." Schurmeier required that engineers plan and record each test on new test data sheets. *Ranger 6* also adopted conformal coating, a method to cover all exposed metal wiring with plastic to preclude short circuits from floating debris. NASA headquarters and JPL delayed the next *Ranger* flight until early 1964 to ensure that the changes took effect.[66]

During 1963, JPL scheduled no launches but concurrently worked on *Ranger 6*, *Surveyor*, and *Mariner Mars 1964* (*MM64*). *MM64* was JPL's next planetary project, planned to launch in November 1964 to fly by Mars. Administrative mechanisms for *Ranger 6* and *MM64* converged during this time, while managers paid relatively little attention to *Surveyor*, which JPL contracted to HAC.[67]

The first test of Ranger's enhanced management was *Ranger 6*'s launch in January 1964. Whereas *Ranger 5* had fourteen significant failures in testing prior to launch, *Ranger 6* had only one subsystem failure during its life cycle testing, boding well for its future. The only hardware that caused trouble was the single new major element, the television camera from Radio Corporation of America (RCA). *Ranger 6*'s launch was flawless, except for an unexplained telemetry dropout. So too were its cruise and midcourse correction. Expectations rose as *Ranger 6* approached its final minutes, when the spacecraft was to take pictures just prior to crashing onto the surface. Reporters, engineers, managers, and scientists waited anxiously in a special room at JPL during the last hour for the first pictures to be broadcast. None ever came. Pickering, who arrived just prior to *Ranger 6*'s impact, was humiliated, saying "I never want to go through an experience like this again—never!"[68]

Investigations ensued, followed this time by congressional hearings. JPL's board isolated when the failure had occurred and which components had failed. Board members could not determine the cause of the failure but nonetheless recommended engineering changes to deal with the several possible causes. NASA Associate Administrator Robert Seamans assigned Earl Hilburn to lead the headquarters investigation of the troubled program. Hilburn's report claimed that the spacecraft had numerous design and testing deficiencies and called for the investigation to be expanded to the entire program. Project personnel vehemently disagreed with the tone and many specific recommendations. JPL incorporated some minor changes but refused to modify its testing practices.[69]

Minnesota congressman Joseph Karth headed the congressional investigation, which focused on JPL's status as a contractor and NASA field center as well as its apparent refusal to take direction from NASA headquarters. Congress recommended that JPL improve its project and laboratory management. Pickering complied by strengthening project management, but he hedged on hiring a laboratory operations manager to assume some of his responsibilities. When NASA Administrator James Webb refused to sign JPL's renewal contract unless it complied, JPL finally hired Maj. General Alvin Luedecke, retired from the air force, as general manager in August 1964. While the successful flight of *Ranger 7* in July 1964 finally vindicated the troubled project, NASA managers became concerned with JPL's third program, Surveyor.[70]

By late 1963, JPL managers spotted trouble signs as Surveyor moved from design to testing. Despite indications of escalating costs and slipping schedules, personnel limitations and preoccupation with Ranger and Mariner prevented JPL managers from adding personnel to Surveyor.[71] However, NASA headquarters managers held a design review in March 1964 to investigate.

The headquarters review uncovered difficulties similar to those on Ranger. Surveyor's procurement staff at JPL was "grossly out of balance" with needs, far too small given that the project consumed one-third of JPL's budget. The review team recommended that both JPL and HAC give project managers more authority over the technical staff. It also recommended that JPL have "free access to all HAC subcontractors," that JPL schedule formal monthly meetings with HAC and the subcontractors, and that JPL's Reliability and Quality-Assurance Office closely evaluate hardware and testing. HAC's management processes also received criticism. The HAC PERT (Program Evaluation and Review Technique) program did not account for all project elements, leading to inaccurate schedule estimates. HAC's change control system, inherited from manufacturing, was unduly cumbersome. Most critically, HAC did not "flag impending technical problems and cost overruns in time for project management to take corrective action."[72]

JPL engineers began their own technical review in April, using twenty experienced Ranger and Mariner engineers. Their primary finding was the inadequacy of HAC's systems engineering. HAC divided the spacecraft's tasks into one hundred discrete units, instead of the eight to ten subsystems typical for JPL. JPL found that many groups "showed a surprising lack of information or interest" about the impact their product had on adjacent products or on the spacecraft as a whole. The spacecraft's design showed HAC had performed few trade-offs between subsystems, leading to a complex design. This led to reliability problems because HAC's design had more components and critical failure points than necessary.[73]

The design reviews came too late to fully compensate for three years of inattention. In April 1964 the first lander "drop test" came to a premature end when the release mechanism failed and the lander crumpled upon ground impact. JPL management responded by increasing JPL's Surveyor staff from under one hundred in June 1964 to five hundred in the fall of 1965. In the meantime, five independent test equipment and spacecraft failures doomed the second drop test in October 1964.[74]

HAC partially complied with NASA recommendations—it strengthened its project organization in August 1964. JPL resisted the headquarters pressure to change, but after some pointed letters from Office of Space Sciences and Applications head Homer Newell, and not-so-subtle pressure from Congress and Administrator Webb, JPL relented and "projectized" Surveyor and its other programs. Both JPL and HAC added personnel and improved budgeting, scheduling, and planning tools. JPL created the Project Engineering Division, which assisted flight projects in "launch vehicle integration, system design and integration, system test and launch operations and environmental requirements."[75]

JPL reassigned engineers from Ranger and Mariner to Surveyor, and it instituted an intense program of contractor penetration. While JPL stated that this led to a clearer picture of problem areas and a better relationship between JPL and HAC, HAC engineers and managers frequently viewed it as an exercise in educating JPL personnel, instead of letting them fix the program's myriad problems. In 1965, HAC estimated that approximately 250 JPL engineers were at HAC's facility on any given day. Numerous changes led to months of intense contract negotiations between the two organizations.[76]

Surveyor's problems caught Congress's attention, leading to a House investigation in 1965. The subsequent report identified "inadequate preparation" and NASA's inattention as the primary problems, resulting in "one of the least orderly and most poorly executed of NASA projects." Congress did not believe further changes to be necessary but severely criticized NASA for its past performance. The investigation concluded, "NASA's management performance in the *Surveyor* project must be judged in the light of a history of too little direction and supervision until recently." Nonetheless, five of Surveyor's seven flights eventually succeeded.[77]

Ranger and Surveyor were JPL's trial by fire. Embarrassing failures and cost and schedule overruns plagued early efforts in both programs. They humbled JPL and provided leverage for Congress and NASA headquarters to impose their will on JPL. JPL managers and engineers learned that they needed strong project management, extraordinary attention to design details and manufacturing, change control, and much better preliminary design work before committing to a project.

When Pickering finally agreed to change JPL and its processes, Jack James's

Mariner project provided the model. JPL survived its early managerial and technical blunders on Ranger and Surveyor primarily because of its solid successes on Mariner. Jack James's strong project organization, backed by progressive design freezes and change control, made Mariner a much different organization than Ranger or Surveyor. Based on the Mariner model, later projects would earn JPL its reputation as the world's leader in the art of deep space exploration.

The Premier Planetary Spacecraft Builder

While Ranger and Surveyor floundered, the Mariner project showed JPL's technical and managerial abilities at their best. After the successful flight of *Mariner 2* (also known as *Mariner R*) past Venus in 1962, JPL targeted *Mariner 3* and *Mariner 4* at Mars, planning to launch during the next opportunity in November 1964.[78] Together, they composed the MM64 project, which extended methods adopted from Mariner R and Ranger.

MM64 manager Jack James used committees established on Ranger to help coordinate across the contributing organizations: JPL, Cape Kennedy, Lockheed, and Lewis Research Center. Four committees coordinated guidance, control and trajectories, tracking and communication, launch operations, and launch vehicle integration. They had no official authority but made recommendations to project management. Headquarters named JPL the project management institution to which the other organizations reported.[79]

James improved communication between the project office, JPL's technical divisions, and external organizations. Mariner managers and engineers extended the concept of the hardware interface to include operational and management interfaces, including the spacecraft, launch vehicle, space flight operations, project and technical division management, science instruments, and operations. James enlisted the cooperation of JPL technical divisions by creating the Project Policy and Requirements document, which served as a "compact between the JPL Project Office and the JPL Line Management for execution of the project." Each project manager met weekly with division representatives to consider "the most serious problems facing his particular area." JPL also added a monthly meeting with division managers to ensure that they

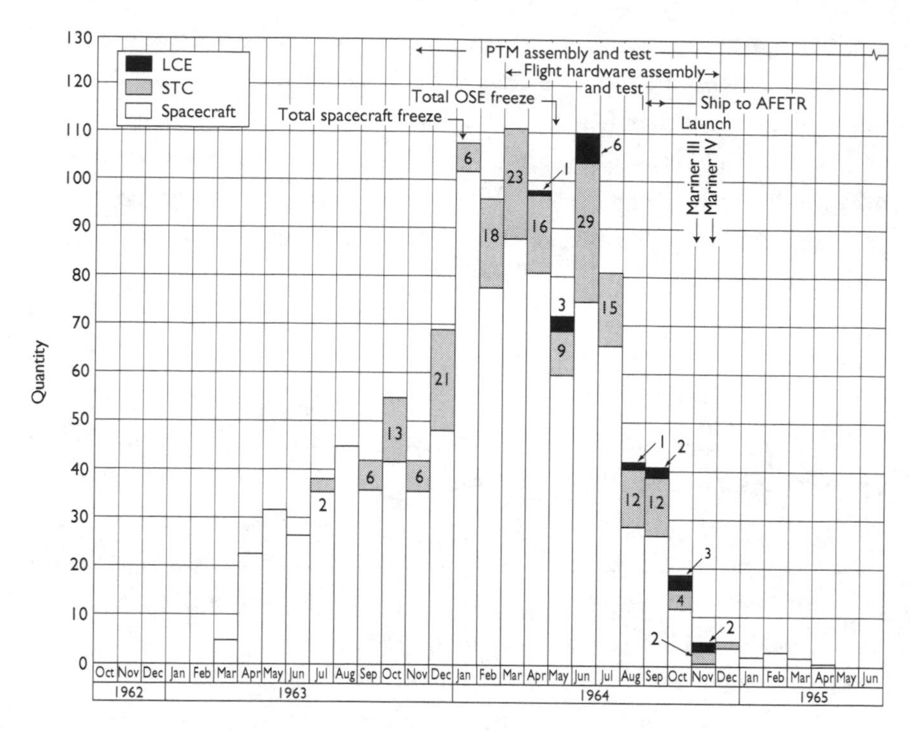

By 1964, JPL learned by experience the typical profile of engineering changes and, consequently, how better to predict costs and schedules, as shown in this change request chart for *Mariner Mars 1964.* Adapted from *From Project Inception through Midcourse Maneuver,* vol. 1 of *Mariner Mars 1964 Project Report: Mission and Spacecraft Development,* Technical Report No. 32-740, 1 March 1965, JPLA 8-28, 32, figure 20.

were familiar with Mariner's problems and that they released personnel to work on them. These measures ensured sufficient attention to Mariner and made JPL's matrix structure work.[80]

Change control had been one of James's innovations on Mariner R, and he formalized it for MM64. The change control system expanded to include progressive "freezing" of specifications and interface drawings as well as hardware, culminating in a final spacecraft design freeze in January 1964 and a support equipment freeze in June 1964. After a freeze, changes could be made only through a change board, which allowed only modifications required for mission success. Project managers kept statistics on changes, noting that the majority of the project's 1,174 changes occurred at subsystem interfaces and in subsystems that contained state-of-the-art equipment.

Project managers also formalized other processes developed first on Mariner R. MM64 management added requirements for parts screening, problem reporting, in-process inspection, comprehensive documentation, and "rigorous status monitoring." The managers continued environmental tests, system tests, and quality assurance procedures developed from Corporal through Ranger. James also continued the Mariner R practice of the "P list." Any problem making the P list received special attention, with "the most effective people available" assigned to solve the problem.[81]

James monitored progress through the use of three sets of schedules and through regular and special reports. The primary schedule reported top-level events and milestones in Gantt (bar) chart format to headquarters. The secondary schedules consisted of Gantt charts for each subsystem, major component, or task. JPL managers called their third set of schedules flow charts, which represented the flow of all of the equipment destined to be integrated in the system test.[82] These network charts "resembled PERT in format and intent" but were "intentionally not so extensive as to require handling by a computer." Network flow charts showed the project's critical path and schedule interactions of all subsystem components, integrated and updated from data supplied by JPL's divisions. The project required updated schedules every other week, in conjunction with a formal report that compared progress with the schedule. Every two weeks, project personnel compiled the data on manual sort-cards that managers manipulated to discern trends and financial implications. Managers monitored some 1,100 flow chart events.[83]

When combined with the experience of JPL's engineering staff, Mariner's organizational techniques ultimately yielded success. *Mariner 3* launched in November 1964, only to be declared dead within nine hours. The problem was in the design of the launch vehicle shroud protecting the spacecraft, designed by NASA's Lewis Research Center. JPL took charge of the investigation and quickly developed a solution, leading to the flawless launch of *Mariner 4* on November 28. Although the spacecraft had some in-flight difficulties, JPL engineers guided the craft to a spectacular conclusion in July 1965, as the spacecraft beamed 21 pictures of Mars back to Earth, as well as analyzing Mars's atmosphere. JPL's success contrasted sharply with five Soviet failures to reach Mars.[84]

Later *Ranger* and *Surveyor* flights confirmed that JPL had dramatically improved its spacecraft management and engineering expertise. *Ranger*'s last two flights, in February and March 1965, were technically superb. Between June 1966 and January 1968, JPL launched seven *Surveyor* spacecraft to land on the Moon, five of which succeeded.[85]

Surveyor's management underwent significant changes late in the project. In September 1966, JPL managers changed the task structure of the HAC contract to a new system known as work package management,[86] which realigned cost accounting and monitoring of tasks "to the individual performing groups in the contractor's organization." Along with the work breakdown structure, JPL required that HAC submit monthly financial reports with more detailed technical, cost, and schedule information. JPL and HAC management met once per month to cover these topics, with a further "consent to ship" meeting scheduled prior to the shipment of each spacecraft to review its test history and problems. HAC and JPL managers developed a thorough "trouble and failure reporting system" that they considered innovative enough to publish a special report on it. The process recorded all test anomalies, required failure analysis by cognizant engineers, involved independent assessments by HAC and JPL organizations, and provided status of failure reports and actions categorized by mission criticality.[87]

The Mariner Venus 1967 program (MV67) further formalized JPL's management and systems engineering. Taking advantage of this, MV67 used the MM64 design as a baseline. Project manager Dan Schneiderman, former spacecraft systems manager for Mariner R and MM64, defined a new management approach at the beginning of the project in a document entitled "Project Policy and Requirements." He froze the entire MM64 design at the outset, requiring change control for any modifications necessary for the Venus mission. The project used three test models: one for antenna and development testing, another for temperature control testing, and the third for flight hardware environmental qualification. Engineers also used the qualification model for simulation and command checking during mission operations. Quality assurance and reliability engineers screened parts, tracked and analyzed failure reports, performed failure mode analyses, verified test procedures, and witnessed tests.[88]

Schneiderman gave the spacecraft system engineer substantial responsibility, including preparation and publication of design specifications books for the flight and test equipment. Subsystem engineers supplied "functional specifications" for their subsystems and support equipment. The spacecraft system engineer also maintained current interface and configuration drawings and mediated "disputes arising out of disagreements between subsystem circuit designers." Change control procedures and subsequent design modification lists were also that engineer's responsibility. The system manager, the spacecraft system engineer's supervisor, ran periodic reviews, which included the spacecraft systems interface and subsystems design review, the spacecraft hardware review, the spacecraft preshipping acceptance review, the launch readiness review, and quarterly headquarters reviews.[89]

MV67's managers and engineers also trained spacecraft operators through the testing process. System testing checked interfaces between subsystems, between the spacecraft and the launch vehicle, and between the spacecraft and the mission operations system and operators. JPL engineers found that they could train mission operators prior to flight by involving them in the integrated system testing at JPL and on the launch pad. The mission operations team members communicated with the spacecraft during these tests using their normal commands and equipment, and they ran compatibility tests with the Deep Space Network. *Mariner 5,* launched in June 1967, arrived at Venus in October. It functioned well, returning data from its atmospheric experiments.[90]

Later JPL managerial innovations included separation of configuration control paperwork and project scheduling from the system engineer. This routine work was given to a separate Project Control and Administration organization. JPL ultimately required all engineering change requests to include cost and schedule impacts along with the technical changes, in effect recreating a version of the air force's configuration management.

JPL went on to pursue new missions to Venus, Mars, and Mercury. Eventually there were the famous Voyager missions to Jupiter, Saturn, Uranus, and Neptune. The laboratory's success showed the maturity of its processes and experience. JPL's preeminence in deep space exploration was undisputed.

Conclusion: The JPL System

From its early beginnings as a student research project, JPL relied on its own expertise. Its engineers developed new technologies prior to and during World War II and contracted their successful solid-rocket innovations to industry. Corporal and Sergeant continued this pattern, with JPL performing the initial analysis, design, and development and contracting to industry for manufacturing. In the NASA era, JPL continued to develop new technologies, contracting for small items that it did not want to manufacture, or as with Surveyor, when it did not have enough personnel to take on more work.

Recognition of the "systems concept" marked JPL's transition from research to engineering development. JPL engineers found that they could not develop entire weapons and their operations using research structures and processes. Engineers had to develop all aspects of the missile, not just those that were "technically sweet." By the mid-1950s, the difficult experience of Corporal led to the systems approach on Sergeant, with formal methods to ensure reliability and operational simplicity. In the late 1950s, JPL reverted to informal processes to create small spacecraft in a great hurry, leading to a spotty reliability record. By the mid-1960s, after disaster on Ranger, JPL engineers and managers had learned once again not to rush into building systems before laying the groundwork.

Reliability was another concept JPL learned from Corporal and Sergeant. Corporal had an abysmal operational record, partly because of the failure of electronic components when shaken by rocket engine vibrations, and partly because of a design never intended for operational use. These two lessons formed JPL's primary belief regarding reliability: good design, solid manufacturing practices, and rigorous testing made a reliable product. This approach served JPL well—but not well enough for deep space. New kinds of failures plagued JPL's early spacecraft, including short circuits caused by floating particles, and software errors. JPL solved these problems through performing component inspection, using simpler designs, coating exposed wires with insulating materials, and instituting "systems tests" to flush out interactions between subsystems and in command sequencing.

Change control became one of JPL's primary means to control projects. Jack James, project manager for Mariner Venus 62 and MM64, developed progressive design freeze on Sergeant to ensure delivery of design information from JPL to Sperry. James and his supervisor, Robert Parks, used the concept again on the Mariner Venus 62 project, then formalized its use on MM64. The Ranger project began to use James's new process after the *Ranger 5* investigation.

Systems engineering, which began as coordination between technical divisions and between JPL and its contractors, became a hallmark of JPL. By 1963, JPL engineers taught space systems engineering at Stanford, where Systems Division Deputy Chief John Small described systems engineering as the "coordination of several engineering disciplines in a single complex effort." According to Small, systems engineers looked at the interfaces and resolved "problems so as to benefit the overall system." They also coordinated the overall test program, defined command sequences to operate the spacecraft, and analyzed "the various interactions" between subsystems to determine where subsystem redundancy would most improve chances for mission success.[91] Other engineers described systems analyses and tradeoffs performed to determine the best mix of components and operations for a given mission.[92]

JPL engineers repeatedly found that many technical problems could be solved only by using organizational means. Problems with missile reliability demanded engineering design changes, parts inspections, and test procedures. Systems engineers solved interface problems by maintaining interface drawings, mediating subsystem disputes, and chairing change control meetings to track and judge design modifications. By 1965, JPL's managers and engineers had learned these lessons well and had become the technical leaders they always believed they were.

JPL developed organizational processes equivalent to those created for the air force's ballistic missile programs. Strong project management, systems engineering, and change control formed the heart of JPL's system, just as they had in Schriever's organization. Both organizations developed them as responses to reliability problems and to political pressures from higher authorities. For JPL and the air force, engineering processes for reliability and change control as well as managerial processes for project and configuration management formed the basis for large-scale development. Although JPL influenced robotic spacecraft development and organization at NASA, it had relatively little influence on NASA's manned programs. The manned programs could have learned from JPL's experiences of the 1950s and early 1960s. Instead, they underwent their own crises. Rather than asking for help from their sister field center, they instead turned to the air force.

FIVE

Organizing the Manned Space Program

> The really significant fallout from the strains, traumas, and endless experimentation of Project Apollo has been of a sociological rather than a technological nature; techniques for directing the massed scores of thousands of minds in a close-knit, mutually enhancive combination of government, university, and private industry.
>
> —T. Alexander, in *Fortune*

By far the largest programs within the National Aeronautics and Space Administration (NASA) during the 1960s were the manned space projects Mercury, Gemini, and Apollo. These differed from other NASA programs because of their massive scale and because several field centers, not just one, contributed significantly to them. The NASA headquarters role was bigger for these huge projects than it was for smaller ones: headquarters coordinated the work of the different field centers. The manned space program contributed disproportionately to the management philosophy and style of NASA as a whole, defined by agency-wide procedures.[1]

While astronauts grabbed public attention, NASA managers and engineers quietly created the machines and procedures necessary for astronauts and ground controllers to operate them. With their personnel descended from German rocket pioneers and National Advisory Committee for Aeronautics (NACA) researchers, NASA's informal groups brought years of aircraft and rocket design expertise to spacecraft design. These new technologies demanded strict attention, and there were the usual number of failures. NASA personnel had to learn how to design manned spacecraft and man-rated rockets as well as how to direct thousands of new employees and scores of contractors.

Difficulties in making the transition from engineers to managers led NASA executives to look elsewhere for people with strong organizational skills. Executives turned primarily to the air force, an organization that developed technologies similar to NASA's. From its inception, NASA had used military personnel, but the importation of experienced air force officers reached its peak in 1964 and 1965, as the newly installed Apollo program director, Brig. Gen. Samuel C. Phillips of the air force, arranged the transfer of scores of air force officers to bring order to NASA's chaotic committees. Phillips imported air force methods such as configuration control, the Program Evaluation and Review Technique (PERT), project management, and Resident Program Offices at contractor locations. By the end of Apollo, Phillips had grafted significant elements of Air Force Systems Command (AFSC) onto NASA's original culture.

Management by Committee, 1958–1962

At its inception in October 1958, NASA consisted of field centers transferred from other organizations. Three centers from NACA formed NASA's core: the Langley Aeronautical Laboratory in Hampton, Virginia; Lewis Research Laboratory in Cleveland, Ohio; and Ames Research Laboratory near Sunnyvale, California. NACA researchers concentrated on empirical and mathematical investigations of aircraft design, including the "X series" of high-performance aircraft, high-speed aerodynamics, jet engines, and rocket propulsion.[2]

An ad hoc group of NACA researchers known as the Space Task Group (STG) promoted the development of space flight. By 1958, they had developed a blunt body capsule design to put a man into space. One other organization, transferred to NASA in January 1960, was key to NASA's manned space efforts: Wernher von Braun's Army Ballistic Missile Agency (ABMA) in Huntsville, Alabama, and its launch facilities at Cape Canaveral, Florida. NASA renamed the ABMA the Marshall Space Flight Center (MSFC), and the Cape Canaveral facilities eventually became the Kennedy Space Center (KSC).[3] These two centers created the massive rockets and launch facilities necessary to place men on the Moon.

The STG's manned capsule, now christened *Mercury*, topped NASA's agenda. Engineers in the Mercury project were to create not only a space capsule but also a worldwide communications network. They were to marry the capsule and the network to the launchers under development by the army and air force. Langley Assistant Director Robert Gilruth headed the STG, which grew quickly and in 1962 moved to Houston, Texas, becoming the Manned Spacecraft Center (MSC).[4]

Gilruth was typical of NASA's experienced engineering researchers. He graduated in 1936 with a master's degree in aeronautical engineering from the University of Minnesota, where he designed high-speed aircraft. From Minneapolis he moved to Langley Research Laboratory, developing quantitative measures for aircraft flying qualities, a job that later served him well in developing manned spacecraft. His *Requirements for Satisfactory Flying Qualities of an Aircraft* became the standard for the field for some years. Gilruth's next assignment brought him back to high-speed aircraft: developing wind tunnel techniques to measure hypersonic flow. Hypersonic flow problems led him to perform full-scale experiments dropping objects from high altitudes at Wallops Island off the Virginia coast. These experiments brought him into contact with rocketry, as he and his NACA colleagues developed and launched rockets to test their theories and technologies.[5]

In the early manned programs, Gilruth treated his engineering colleagues as technical equals. As his assistant, Paul Purser, described it, "Individuals around the conference table are not aware of being division chiefs or section heads—they are all people working on a problem." Gilruth's ability and experience made him more than just a manager. Future NASA Administrator George Low said in the early 1960s, "Gilruth works personally with many people in the Space Task Group. His method of operation is one of very close technical involvement in the project. He could tell you . . . every nut and bolt in the Mercury capsule, how it works, and why it works. I've been in many meetings over the last two or three years, where the whole picture would look very complex. After perhaps a half-hour's discussion, Gilruth would come up with the right solution, and the rest of us present would wonder why we hadn't thought of it."[6] The STG's hands-on approach to engineering would continue for years to come.[7]

In the fall of 1958, the STG established Mercury's basic configurations and missions. Engineers planned to use existing rockets to launch the new spacecraft—first von Braun's proven *Redstone* and *Jupiter* boosters, then the air force's more powerful but less mature *Atlas.* Congress gave NASA the same procurement regulations as the Department of Defense (DOD). Thus the new organization held a bidder's conference in November 1958 to describe the proposed system to contractors, mailed out bid specifications, and required responses in 30 days. In January 1959 NASA awarded the *Mercury* spacecraft contract to McDonnell Aircraft Corporation, a cost-plus-fixed-fee contract for $18,300,000 and an award fee of $1,150,000. STG engineers also started negotiations with the army and the air force for launchers.[8]

Two traditions distinguished Mercury's management: the informal structure and procedures of the STG, and the more formal approaches of McDonnell Aircraft and the air force. STG engineers and scientists used committees characteristic of research, simply creating more of them as Mercury grew. McDonnell Aircraft brought structured methods developed from years of interaction with the air force. Engineers from the STG and McDonnell Aircraft worked closely to resolve the numerous novel problems they encountered. For Atlas, the air force used its own procedures and supplied representatives to STG committees to define interfaces between *Atlas* and the *Mercury* capsule.[9]

Mercury's driving force was a "bond of mutual purpose": determination to regain national prestige, fear of Soviet technical accomplishments, and pride in American capabilities. Managers gave tasks and the authority to perform them to young engineers. As one STG veteran put it, "NASA responsibilities were delegated to the people and they, who didn't know how to do these things, were expected to go find out how to do it and do it."[10] Working teams phased in and out as they completed their tasks; frequently they determined "a course of action and proceeded without further delay, with verification documentation following through regular channels." In other words, engineers took immediate action without management review and left others to clean up the paperwork.[11]

This worked because of the extraordinary "flatness" and open communication prevalent throughout the organization. STG leaders insisted that problems be brought into the open. According to Chris Kraft, later the director of Johnson Space Center, all the people in the organization felt as if they could say "what they wanted to say any time they wanted to say it"; Kraft called

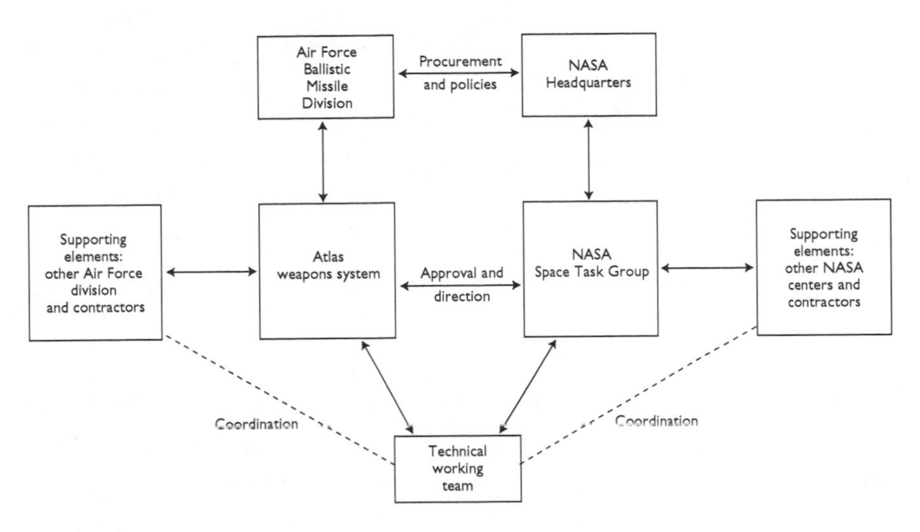

The Mercury-Atlas organization was extraordinarily flat, with only three levels from NASA and air force headquarters to the working groups who built the spacecraft and launchers. Adapted from *Mercury Project Summary, Including Results of the Fourth Manned Orbital Flight,* SP-45 (Washington, D.C.: NASA, 1963), 19, figure 1-8.

this STG's "heritage."[12] Managers tracked events through informal communications and frequent technical reviews. They directed resources to problem areas, but they seldom intervened.[13]

On Mercury, and later on Gemini and Apollo, this sense of common purpose also prevented the development of bureaucratic sclerosis. As stated by one of the engineers on Gemini and Apollo, "You would see people who would try to build empires, who would try to be obstructionist, and they would be just absolutely steamrolled by this team. I saw it time and time again where there was this intense feeling of teamwork. It wasn't always smooth, but it was like, 'We've got a common goal.'"[14]

The STG's multiple committees became unwieldy as Mercury grew from a nucleus of 35 people in October 1958 to 350 in July 1959. To deal with the proliferation of committees, STG created another committee, the Capsule-Coordination Panel, subsequently upgraded to an office in Washington, D.C.[15]

NASA headquarters executives soon realized that the STG and McDonnell had drastically underestimated the scope, cost, and schedule of the program. Within two months of its beginning, the estimated cost of the McDonnell contract was $41 million, more than twice the initial estimate, while the air force's estimated costs for the *Atlas* boosters increased from $2.5 million to $3.3 million. These increases led to a round of cost-cutting measures, yet costs continued to rise. The estimated cost of the McDonnell contract reached $70 million by January 1960. NASA Administrator Keith Glennan's initial response was to visit the STG in May 1959. He came away impressed by the esprit de corps in the STG and the size and complexity of the project. With Congress and the administration willing to foot the bill and the STG rapidly tackling technical issues, Glennan elected not to intervene.[16]

In the summer of 1959, Gilruth organized the New Projects Panel to identify manned projects beyond Mercury. The panel identified circumlunar flight (not landing) as the most promising goal.[17] The most important technical developments for a manned Moon mission were in the military's rocket and engine programs. In January 1959, NASA acquired the air force contract with North American Aviation (NAA) to develop the huge liquid-fueled *F-1* engine. NASA acquired the *Saturn I* launcher in January 1960 along with von Braun's rocket team. *Saturn* was the only launch vehicle then under development that promised sufficient size for a manned lunar landing program. However, for the moment, it was a launch vehicle without a mission.[18]

In early 1960, NASA's advanced planning groups concluded that a lunar mission was the best next step. NASA named the proposed new program Apollo, and in August managers announced they would award three contracts to industry for feasibility studies. The STG selected the Martin Company, General Electric (GE), and the General Dynamics Convair Division to perform the studies. Study guidelines were so vague that when the Martin Company engineers reported back in December 1960, STG engineers told them to include astronauts and to consider lunar landing and recovery. MSFC also sponsored its own feasibility studies, while the STG started an internal study.[19]

After the Bay of Pigs disaster and Yuri Gagarin's flight in April 1961, the Kennedy administration proved receptive to NASA's lunar mission planning. On May 25, President John F. Kennedy proposed that NASA land a man on the Moon "before the decade is out." Congress enthusiastically agreed and immediately increased NASA's funding.[20]

STG managers quickly moved Apollo from feasibility studies to development. Gilruth had prepared the groundwork, creating the Apollo Project Office in September 1960. Although the hardware configuration remained uncertain, the STG forged ahead, dividing the system into six contracts: launch vehicles, spacecraft command module and return vehicle, propulsion module, lunar landing stage, communications and tracking network, and launch facilities. Just before selecting NAA for the spacecraft command module in November 1961, MSC engineers changed the Statement of Work, meaning that they awarded NAA a contract to build a command module based on specifications that NAA's managers and engineers had never seen.[21]

With four Saturn stages under development, von Braun's MSFC engineers did not have the resources to design, manufacture, and test all of the vehicles. They had to rely upon industry instead of their traditional in-house design. MSFC managers transferred *S-I* stage development and manufacturing to Chrysler and awarded the new *J-2* cryogenic engine to NAA Rocketdyne in June 1960.[22]

MSFC inherited strong technical divisions from its army heritage, each based on specific disciplines such as rocket propulsion, structures, or avionics. The technical divisions coordinated project work through committees, contributing to MSFC's typically large, interminable meetings. Contractors complained that MSFC managed by technical takeover. However, under the pressure of having many large, complex projects, project and matrix management made inroads into MSFC's discipline-based, functional organization. The divisions fought the change, forcing MSFC Director Wernher von Braun to clarify the power of project managers vis-à-vis the technical divisions.[23]

Von Braun was one of the world's leading rocket engineers and a charismatic leader. Even when immersed in administrative duties, he closely followed the technical details of MSFC's rockets. Von Braun secured inputs from all participating engineers and technicians and arrived at a consensus through group meetings. He used an informal but disciplined system of weekly notes, requiring subordinates *two* levels below him to send him one page of notes summarizing the week's events and issues. Von Braun then wrote comments in the margins of these notes, copied the entire week's set, and distributed the notes for everyone in MSFC to read. They became popular reading because they contained the boss's detailed comments on MSFC events and people.

This system of "Monday Notes" had a number of important ramifications. First, because von Braun required the notes to come from two levels below, the managers directly under him could not edit the news he received. Second, because of having to send weekly notes to von Braun, all managers formed their own information-gathering mechanisms. Third, the redistributed notes with von Braun's marginalia provided a mechanism for cross-division information flow, because everyone saw comments on not only their own activities but the activities of all other divisions. The notes moved information vertically—from the managers and engineers up to von Braun, and from von Braun down to the managers and engineers—as well as horizontally—from division to division.

This very open communication system provided MSFC engineers and managers with advanced notice of potential problems, often spurring critical problem-solving efforts across the divisions. Some MSFC engineers complained about the extraordinary communication technique because it created an "almost iron-like discipline of organizational communication" in which "nobody at the bottom really felt free to do anything unless he got it approved from the next level up, the next level up, the next level up." However, it did ensure that information flowed quickly and effectively throughout the organization.[24]

Von Braun required all MSFC personnel to take "automatic responsibility" for problems. If MSFC employees found a problem, they were to solve it, find someone who could solve it, or bring it to management's attention, whether or not the problem was in their normal area of responsibility. This intentional blurring of organizational lines helped create an organization more interested in solving problems than in fighting for bureaucratic turf.[25]

Apollo planners soon recognized a gap between Mercury's short flights and the long flights and complex operations of Apollo. To bridge this gap, STG chief Gilruth authorized the Gemini program, which was to modify *Mercury* to accommodate two astronauts, perform orbital maneuvers, and rendezvous with other spacecraft. NASA awarded the *Gemini* capsule contract to McDonnell without competition because it was a modification of *Mercury.* Gilruth split the engineering staff between Mercury and Gemini, and in January 1962 he established the Mercury, Gemini, and Apollo Project Offices.[26]

Like Mercury and Apollo, Gemini used coordination panels for day-to-day management. The Project Office established six panels: three for the spacecraft—mechanical systems, electrical systems, and flight operations—and one each for the paraglider,[27] *Atlas-Agena,* and *Titan II.* They typically held weekly meetings, while the air force used its standard procedures to manage its portions of the program. Because the air force provided the *Titan II* and the *Agena* target boosted on an *Atlas* missile, air force and NASA managers established an additional panel to coordinate between them. Assistant Secretary of the Air Force for Research and Development Brockway McMillan and NASA's Robert Seamans were the co-chairmen, with D. Brainerd Holmes of the Office of Manned Space Flight (OMSF) and Gen. Bernard Schriever of AFSC the highest-ranking members.[28]

Committees coordinated between engineers and managers at headquarters, MSFC, and MSC, particularly for interface designs and characteristics. By July 1963, there were so many committees that Holmes created another one, the Panel Review Board, to coordinate them.[29]

In the white heat of the early post-*Sputnik* era, technical achievement was the primary gauge of space program success, and political leaders left control in the hands of the engineers who promised technical success. Engineers and scientists from the STG and MSFC used committees to coordinate their work, a habit inherited from research traditions of NACA laboratories and von Braun's "Rocket Team." Engineers rapidly developed rockets and spacecraft, with little heed for cost. NASA and its contractors rushed into contracts and designs without firm requirements or a clearly defined mission, making schedule or cost predictions virtually impossible. For example, MSC and MSFC engineers wrote definitive specifications for their Apollo elements well after contract awards—and in the case of MSFC's *Saturn* stage I, long after completion of the initial design, manufacturing, and testing.[30] For the moment, this was not a problem, because Congress gave NASA more money than NASA asked for, allowing a continuation of conservative design traditions on a much larger scale.

Conservative Engineering at the State of the Art

MSFC's conservative tradition of rocket development traced back to von Braun's work with the German Army in World War II and the U.S. Army thereafter. Step-by-step, precise methods characterized von Braun's approach to rocketry. Each rocket drew upon the designs and experiences of previous rockets. Engineers made small changes in each successive rocket and tested each change to ensure that it did not compromise the design. They considered flights successful even if they ended in explosion as long as they collected sufficient data to determine the explosion's cause.[31] Once designers found the problem, they modified the design and flew it to test the fix. Based on their many years of experience, they believed it virtually impossible to design an engine or a rocket without significant difficulties.

Although MSC and MSFC engineers differed in a number of ways, one similarity was distrust of contractors. Essentially, both MSFC and MSC engineers took apart and rebuilt the stages and capsules that contractors delivered to them. Boeing's contract for the *Saturn S-I* first stage was a good example. MSFC's original contract, awarded after a competition in 1962, gave Boeing partial responsibility for stage design and assembly but little responsibility for booster specifications or testing. Later contracts gave Boeing progressively more responsibility, until it was responsible for all aspects of the stage except for mission operations.[32]

Borrowing and extending practices from the air force and the navy, NASA closely controlled the contractors. NASA used Resident Manager's Offices to monitor the contractors. NASA realized that on-site surveillance was "somewhat sensitive from the point of view of the contractor" but persisted because of its belief in face-to-face communication, its distrust of contractor capabilities, and its trust in its own capacities. NASA's infringement on contractor prerogatives included forced renegotiation of contracts and designs. For example, Lunar Excursion Module (LEM) contract winner Grumman supposed that it would build the design presented in its bid. Instead, NASA engineers redesigned the LEM and the contract.[33]

NASA Administrator James Webb expanded contractor penetration to penetration of its own field centers. He wanted a separate information channel to check his own organization, and he hired GE for this purpose. The GE Policy Review Board, established in December 1962, was to provide systemwide coordination and integration. Neither MSC personnel nor MSFC personnel wanted headquarters or GE to integrate them, however. As one GE manager later explained, "Frankly, they didn't want us. There were two things against us down there [at MSC]. No. 1, it was a Headquarters contract, and it was decreed that the Centers shall use GE for certain things; and [No. 2] they considered us Headquarters spies."[34] MSC management hampered GE's attempts to fulfill its integration role, prohibiting "unannounced visits" and forbidding GE from taking any significant action unless approved by MSC. Field center resistance was effective, for in July 1963, Apollo's new Panel Review Board abolished GE's board. Later, after a long briefing to Administrator Webb, Apollo program director Phillips removed systems integration from GE's contract.[35]

Headquarters also contracted with American Telephone and Telegraph (AT&T) to provide technical assistance. AT&T created a separate company called Bellcomm to provide headquarters with technical consultants, whom headquarters used to cross-check the field centers in the way that The Aerospace Corporation checked contractors for the air force. Bellcomm ultimately performed exemplary work in trajectory analysis, but field center engineers and managers avoided Bellcomm representatives because they believed them to be headquarters spies. The use of GE and Bellcomm failed to provide NASA executives with the means to control the field centers, because of the unequal power between the government and industry. NASA provided the funds, and industrial contractors were uncomfortable criticizing their source of funding. They knew better than to bite the hand that fed them.[36]

Perhaps the greatest challenge to NASA and contractor engineers was to make their new vehicles absolutely reliable so that humans could fly in them. Redundancy was one common technique to improve reliability. For example, MSFC engineers designed each *Saturn* stage with clusters of engines, so that if one failed, the remaining engines could continue the mission. Man-rating the air force's *Atlas* and *Titan* missiles primarily meant adding redundant electronic circuitry and failure detection circuits to the missile. NASA engineers thought of astronauts as redundant design elements that could improve the chances for mission success by taking over spacecraft functions if components failed.[37]

Ensuring high quality and safety was as much a function of management and organization as engineering design. Because a number of booster and capsule components were single string—meaning that if they failed, there were no separate backups—STG engineers rigorously verified individual components for high quality. In turn, quality depended upon every worker building each component with the finest workmanship.

The manned programs instituted special methods to make factory workers aware of the importance of high quality. Taking advantage of the projects' prestige and visibility, Mercury and Gemini managers distinguished NASA components and workers using special symbols. On the Gemini Titan II program, Martin management gave special worker certifications to top employees, along with orientations, emblems, labels, badges, and even distinctive toolboxes, "painted Air Force blue and individualized with each worker's name." Astronauts visited production lines to encourage high standards and workmanship. Based upon Space Technology Laboratories (STL) recommendations, the program adopted strict inspections and random part checking for quality control. Tight control over manufacturing contrasted sharply with MSC's loose internal structure and processes.[38]

Worker motivation for Apollo was extraordinary, contributing significantly to the high quality of components that went into the project. Most of the time, this was a major advantage. However, occasionally, it caused problems. For example, at NAA, the wiring harnesses for the command module occasionally disconnected when the pins broke off in the connectors. They "found one lady out there who had evidently extremely strong hands, and she knew that she was working on the *Apollo* Program, so when she crimped, she crimped extremely hard, and she could actually crimp hard enough to deform the tool and squeeze the wires to where they were almost broken. She was just trying to do her job a little better than normal, but actually she was causing us a lot of trouble. For her, they put a spatial stop on the tool so that she couldn't crimp it any harder."[39]

Mercury and Gemini engineers created a system to track individual parts with manufacturing and test histories — to ensure that only fully documented, flawless parts became flight components. General Dynamics and Martin managers designated *Mercury* and *Gemini* launch vehicle components as critical, with special tags, paperwork, and procedures. At Martin, vehicle chaperones accompanied each vehicle through manufacturing, moving the vehicle and its paperwork through the factory. Using Martin's program as a model, the air force and NASA initiated a similar program at Lockheed for the *Agena* target vehicle. At the Factory Roll-Out Inspection, NASA thoroughly reviewed records for each vehicle component. Mercury engineers started their quality control program late, resulting in many component replacements when the capsule or its components failed acceptance tests. Learning from this, Gemini managers began their quality control program right at the start.[40]

Military models were the basis for most of the STG's few formal processes early in the program. The Source Evaluation Board was one example, as was the Mock-Up Inspection Board, a formal inspection of a full-scale system model. Another was the Development Engineering Inspection, where STG engineers certified the design to be ready for flight. Hardware qualification was another military import, using rigorous environmental tests to stress components. On the *Apollo* spacecraft, contractors prepared for the Spacecraft Assessment Review, Customer Acceptance Readiness Review, and Flight Readiness Review. Contractors submitted written reports, which NASA committees reviewed. Starting on Mercury, NASA also held flight safety reviews for each spacecraft to review modifications, testing, and preparations for launch and operations.[41]

Engineers at both centers believed in uncovering design problems through extensive tests and analyses. STG Director Gilruth and MSFC Director von Braun both believed that testing led to better understanding and improved designs. MSFC engineers tested engine stability by exploding small bombs in the rocket's exhaust path to ensure that unexpected flow variations in the engine's hot gases would not create instabilities that could lead to an explosion. STG engineers injected electrical failures into their designs to ensure that they would survive. Following air force ballistic missile practices, they searched for critical weaknesses in the design, making changes when they found them.[42]

NASA and its contractors tested components to ensure that they could survive launch vehicle vibrations, the thermal and vacuum environment of space, and electromagnetic interference between electronic components. They also performed life tests to see how long components would last. Once components completed component tests and Preinstallation Acceptance Tests, engineers and technicians integrated them into a capsule. Contractors then ran Capsule System Tests, which verified that electronic wiring was working and that mechanical devices were functioning properly.[43]

One item not easily tested was the performance of the environmental and thermal control system of the entire *Mercury* capsule. Although vacuum chambers existed, none were large enough to test the whole vehicle. In late

1960, STG and McDonnell engineers developed a large vacuum chamber in which they could test the entire capsule's environmental and thermal controls. This task, known as Project Orbit, found numerous problems, particularly in the reaction control system used to keep the capsule pointed correctly in space.[44]

To manage the large number of engineering alterations, the STG established for Mercury a Change Control Group, whose membership fluctuated based on the nature of the problem. Configuration control crept into the program through the air force's supply of launchers for NASA. The air force established configuration control for Gemini's Titan II in December 1962. Gemini engineers "froze" portions of the spacecraft design as early as March 1962, and Apollo contractors froze elements of their designs by June 1963.[45]

Despite these efforts, *Mercury*'s first test flights suffered their share of failures. The first occurred in July 1960 during the first full-scale test of the *Mercury* capsule mounted on an *Atlas* booster. About one minute after launch, the booster failed as it accelerated through the period of highest aerodynamic pressure, leaving NASA and the air force to search for debris in the Atlantic off Cape Canaveral. The evidence pointed to a structural failure in the interface between the *Atlas* and the *Mercury* capsule, based on mechanical differences between the *Mercury* capsule and the *Atlas*'s normal complement of nuclear warheads. Engineers had not found this problem in earlier tests because they could not fully simulate the physical dynamics of the vehicle in flight. Following the failure, the Mercury-Atlas coordination panel formed a new committee to find and resolve interface problems. The next successful test flight, in February 1961, featured a structure-stiffening "belly band." Engineers later included vibration and structural resonance characteristics of the combined launch vehicle–payload system in interface designs and documentation.[46]

Interface problems also dogged a test flight aboard MSFC's *Redstone* booster in November 1960. The launcher lifted off the ground about four inches, then settled back onto the pad while the capsule's escape rocket launched, dragging erroneously opened parachutes. Subsequent investigations traced the failure to a timing problem between the *Redstone* electronics and the launch complex, caused by the different weight of the *Mercury* capsule compared to *Atlas*'s normal payload of weapons. The weight difference caused a slightly delayed mechanical disconnection time for a shutoff signal to the

Redstone engines, resulting in the firing of the capsule escape system. MSFC engineers quickly fixed the problem, and the next test flight, in December 1960, was a success.[47]

These failures caused STG engineers to reevaluate their design philosophy. In a report to headquarters, they stated, "It has become obvious that the complexity of the capsule and the booster automatic system is compounded during the integration of the systems."[48] Engineers from the STG, the air force, and the contractors investigated the causes of the numerous failures and frantically improved the design of the boosters, the capsules, and the interfaces between them. One important way in which NASA and the air force formalized these relationships and processes was through the development of interface specifications that defined electrical and mechanical connections between the capsule and the launch vehicle, along with vibration and acceleration loads.[49]

Mercury flight failures led to increased attention to *Gemini* and *Apollo* interfaces. In June 1963, the air force, NASA, Martin, and McDonnell initiated investigations of the structural and electronic interfaces between *Titan* and the *Gemini* spacecraft. Gemini managers created the Systems Integration Office in February 1964 to monitor spacecraft weight and interfaces between the spacecraft, the launch vehicle, and the *Agena* target vehicle with regular coordination meetings known as Interface Control Panels. Interface Specification and Control Documents recorded the results of these meetings. Engineers developed new tests, including electronic-electrical interference tests, joint combined systems tests, flight configuration mode tests, and wet mock simulated launch tests, all of which verified interfaces and launch procedures on the launch pad. Apollo personnel began to work on interface problems in July 1963, when the Panel Review Board standardized Interface Control Documents and made MSFC the Apollo document repository.[50]

The STG and its successor, MSC, evolved from NACA's hands-on research culture. Despite massive growth, the STG maintained informal mechanisms for coordination and control. MSFC too held fast to its discipline-based army heritage. Both organizations benefited from military and industrial processes used by their contractors. Because of their prestige and deep pockets, the manned programs commanded attention from everyone, from the top of the management hierarchy to the shop floor workers at contractor facilities.

Both MSC and MSFC were ultraconservative concerning their own designs. They emphasized reliability and safety above cost, allowing costs to increase. The exemplary flight records of Mercury and Gemini showed that this conservatism could help ensure that technical objectives were achieved. Considering that the air force was then instituting a program to improve *Atlas*'s reliability to 75%, achieving a perfect flight record was no easy feat.[51] However, the field centers' proclivity to increase costs was a weakness soon exploited by Congress and NASA headquarters to tighten control.

The Cost Crisis of 1962–1964

In technical terms, NASA's organization and procedures proved successful on Mercury and Gemini. Both programs boasted enviable flight records, completing their objectives and successfully returning their astronauts on every flight. On the other hand, both programs featured large cost increases. McDonnell's *Mercury* capsule contract grew at an astounding rate, from an initial contract of $19,450,000 in 1959 to a total of $143,413,000 by the program's end in 1963. Within months of its start, Gemini was headed for large cost increases as well. NASA funded the increases, although Congress began to ask questions in the fall of 1962.[52]

OMSF's Holmes, overseeing the three manned space flight programs, saw the escalating cost trends as well as anyone. A hard-nosed project manager from Radio Corporation of America's (RCA's) Ballistic Missile Early Warning System, Holmes realized that something had to be done if Apollo was to meet Kennedy's 1969 deadline to land a man on the Moon. Holmes also recognized that Kennedy had staked his reputation on the manned program and would go to great lengths to ensure its success. Based on this, he felt no particular need to control costs and instead requested more funds from NASA Administrator Webb.[53]

Holmes first asked Webb to go back to Congress for a supplemental appropriation. Despite Kennedy's support, Webb recognized that Congress was becoming uneasy about NASA's burgeoning costs, and he refused. Holmes next demanded that Webb strip other NASA programs to support the manned program. When Webb again refused, Holmes went over his head, appealing directly to President Kennedy. This infuriated Webb, now placed in the uncomfortable position of justifying why he should not strip other NASA projects to fund Kennedy's priority program. Although Kennedy was not sure that he "saw eye-to-eye" with Webb on this issue, he backed his chosen administrator. Webb replaced his insubordinate OMSF director with STL executive George Mueller (pronounced "Miller").[54]

Under the circumstances, further cost pressures were unwelcome. In March 1963, Gemini project manager James Chamberlin admitted that the project required another huge infusion of money. The unwelcome news led to his replacement by Charles Mathews. Mathews immediately set up a committee to improve cost estimation at NASA and McDonnell. After another year of strained budgets and organizational crises—invariably related to novel portions of the project—Gemini's budget stabilized, leading to an enviable record of flight success. However, cost estimates had risen from $531,000,000 to $1,354,000,000 between 1962 and 1964.[55]

With Congress increasingly concerned with costs, Webb instigated an internal investigation of NASA projects in early 1964.[56] Deputy Associate Administrator Earl Hilburn chaired the investigation to study NASA's project scheduling and cost estimating methods. In confidential reports submitted in September and December of 1964, Hilburn described NASA's abysmal record on cost and schedule control. Engine development showed by far the largest schedule slips, at 4.75 times the original estimate. Hilburn found significant slips in launch vehicles requiring new engines (3.09), manned spacecraft (2.86), "simple" scientific satellites (2.85), and astronomical observatories (2.83). Classified by field center, he found that the Jet Propulsion Laboratory (JPL) had the smallest schedule slips (1.6) and MSC the largest (3.06). The newest centers, MSC and Goddard, showed the largest slips, and the older centers (JPL, Langley) showed the smallest. Costs varied similarly.[57]

Hilburn noted that the DOD experience was also "one of over-runs and schedule slippages in the majority of cases." He concluded that these results were behind recent changes in DOD procurement, "including program definition, and incentive contracting." Based on DOD precedents, NASA had already begun converting contracts from cost-plus-fixed-fee contracts to incentive contracts that awarded firms a higher fee when they performed well and a lower fee when they did not. The Hilburn Report lent support to contract conversion at NASA, which NASA implemented from 1964 through 1966.[58] NASA also adopted the DOD practice of phased planning, requiring contractors to better predict costs and schedules during a definition phase of development. The directive had little initial impact because of disagreements about implementation, and more importantly, because few new projects started in 1966 and 1967.[59]

NASA's manned space programs were among the worst offenders as NASA's budget exploded between 1962 and 1964. Having grown accustomed to generous, even extravagant budget increases without justification or congressional concern, NASA's continued cost excesses were understandable, if not justifiable. As in other organizations, the time came when its initial growth surge and justification had to slow and managers had to predict and hold to funding profiles. At the highest levels, NASA could replace recalcitrant executives and implement new standards and processes, but to truly predict and control costs, NASA would have to implement rigorous procedures on its research and development (R&D) programs.

Systems Management, Air Force Style

The informal management methods that characterized the manned program's first few years did not last. New OMSF chief George Mueller understood very well that he would have to implement rigorous cost prediction and control methods to address NASA's unruly engineers and R&D projects. To do this, he would turn to the management methods with which he was familiar, those developed in the air force and at Thompson-Ramo-Wooldridge (TRW).

Mueller began his career in 1940 working on microwave experiments at Bell Telephone Laboratories. After the war he taught electrical engineering at Ohio State University, and in 1957 he joined Ramo-Wooldridge's STL as the director of the Electronics Laboratories. He moved up quickly, becoming program manager of the Able space program, vice president of Space Systems Management, and then vice president for R&D. Mueller helped build the air force's bureaucratic system to control large missile and space projects.[60]

Before Mueller started his NASA duties, he performed his own investigation of OMSF. His first impression of NASA was that "there wasn't any management system in existence." The many interface committees and panels worked reasonably well but did not penetrate "far enough to really be an effective tool in integrating the entire vehicle." Most seriously, Mueller found no means to determine and control the hardware configuration, leaving no way to determine costs or schedules. Mueller concluded that he had to "teach people what was involved in doing program control."[61]

In August 1963 Mueller invited each of the field center directors to visit him. He explained his proposed changes and how they would help solve NASA's problems with the Bureau of the Budget and the President's Science Advisory Committee. Mueller explained, "If we didn't work together, we were sure going to be hung apart." Mueller had little trouble convincing *Apollo* spacecraft manager Joseph Shea at MSC that his changes were necessary, given Shea's prior experience at TRW. However, he met resistance from MSFC leaders. When MSFC manager Eberhard Rees challenged Mueller's proposals, Mueller retorted that "Marshall was going to have to change its whole mode of operation." Two weeks later, at another MSFC meeting, von Braun gave Mueller "one of his impassioned speeches about how you can't change the basic organization of Marshall." Refusing to back down, Mueller told him that MSFC's laboratories "were going to have to become a support to the program offices or else" they "weren't going to get there from here." Von Braun, in response, reorganized MSFC on September 1, 1963, to strengthen project organizations through the creation of the Industrial Operations branch.[62]

Webb strengthened Mueller's position by making the directors and projects at MSC, MSFC, and KSC report to OMSF. Mueller reduced attendance at the Manned Space Flight Management Council to himself and the center directors to ensure coordinated responses to OMSF problems. Borrowing from the air force Minuteman program, Mueller also formed the Apollo Executive Group, which consisted of Mueller and Apollo contractor presidents. The group members met periodically at NASA facilities, where Mueller pressured them to resolve problems.[63]

Facing the same dilemma Holmes faced—Apollo's slipping schedules and cost overruns—Mueller concluded that the only way to achieve the lunar landing before 1970 within political and budget constraints was to reduce the number of flights. In his "all-up testing" concept, each flight used the full Apollo flight configuration. This approach, used on the Titan II and Minuteman programs, violated von Braun's conservative engineering principles. Von Braun's existing plan used a live *Saturn* first stage with dummy upper stages for the first test. The second test included a live second stage with a dummy third stage, and so on. By contrast, the all-up concept used all flight stages on the very first test. This reduced the number of test flights and eliminated different vehicle configurations, with their attendant differences in designs, ground equipment, and procedures.[64]

Saturn V program manager Arthur Rudolf cornered Associate Administrator Robert Seamans at a meeting, showing him a model of the *Saturn V* dwarfing a *Minuteman* model, saying, "Now really, Bob!" Seamans got the hint—Mueller's concept did not apply to the more complex *Saturn V.* Encouraged, Rudolph showed the same models to Mueller. Mueller replied, "So what?" The all-up concept prevailed.[65]

In November 1963, Mueller reorganized the Gemini and Apollo Program Offices, creating a "five-box" structure at headquarters and the field centers. The new structure (see figure) ensured that the field centers replicated Mueller's concept of systems management and provided Mueller with better program surveillance. Inside these "GEM boxes,"[66] managers and engineers communicated directly with their functional counterparts at headquarters and other field centers, bypassing the field centers' normal chain of command. As one NASA manager put it, "Anywhere you wanted to go within the organization there was a counterpart whether you knew him or not. Whether you had ever met the man, you knew that if you called that box, he had the same kind of responsibility and you could talk to him and get communication going."[67]

Mueller's new organization initially wreaked havoc at NASA headquarters, because the change converted NASA engineers who monitored specific hardware projects into executive managers responsible for policy, administration, and finance. For several months after the change, headquarters was in turmoil as the staff learned to become executives.[68]

NASA's organizational structure changed as a result of Mueller's initiatives, but he could not always find personnel with the management skills he desired. Shortly after assuming office, Mueller wrote to Webb, stating that NASA could use military personnel trained in program control. Because of the air force's interest in a "future military role in space," Mueller believed that it would agree to place key personnel in NASA, where they would acquire experience in space program management and technology. NASA, in turn, would bene-

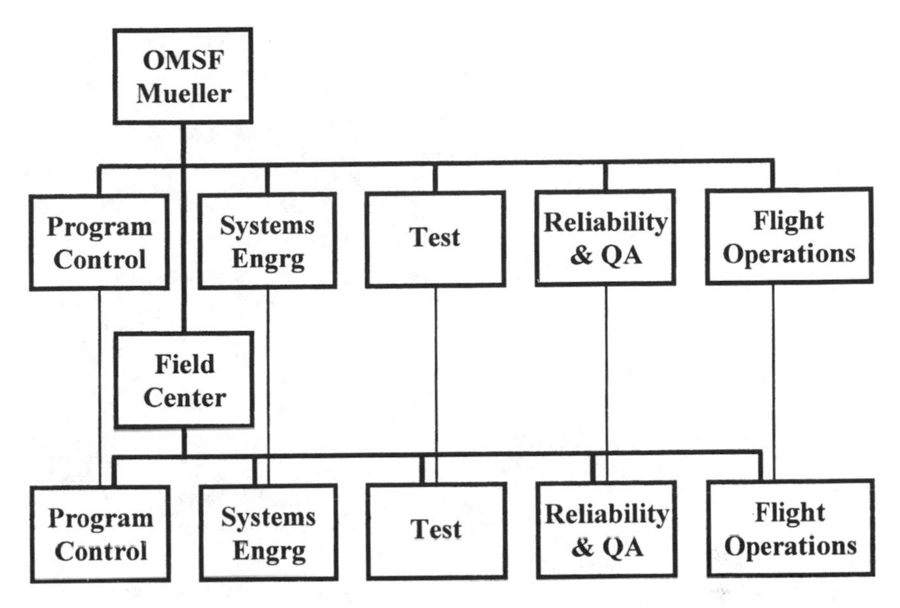

George Mueller's "five box" structure replicated the headquarters OMSF organization through the OMSF field centers. Adapted from "Miscellaneous Viewgraphs," circa January–February 1964, "Organization and Management," folder LC/SPP-43:11.

fit from their contractual experience and program control methods. Mueller stated, "It is particularly worth noting that the Air Force, over a period of years, has developed the capability of managing and controlling the very contractors upon whom we have placed our primary dependence for the lunar program." He proposed to place Minuteman program director Phillips in the position of program controller for OMSF and contacted AFSC chief Schriever regarding this assignment. Schriever agreed, but only on the condition that Phillips become Apollo program director. From this position, Phillips would transform NASA's organization and would become known as Apollo's Rock of Gibraltar.[69]

Phillips surmised, "NASA had developed to be a very, very professional technical organization, but they had almost no management capability nor experience in planning and managing large programs."[70] Phillips turned to the air force for reinforcements and to his most valuable tool from Minuteman, configuration management.

In a January 1964 letter to Schriever, Phillips asked for further air force per-

George Mueller (left) and Samuel Phillips (right) imposed air force management methods on Apollo by introducing new procedures and bringing dozens of air force officers into NASA's manned space flight programs. Courtesy NASA.

sonnel to man OMSF's program control positions. After AFSC assigned two officers to NASA, Phillips created a list of fifty-five positions that he wanted to fill with air force officers. Such a large request entailed formal negotiations between NASA and the air force for Phillips's "Project 55." Secretary of the Air Force Eugene Zuckert agreed to consider transfers if NASA better defined the position requirements, so that he could ensure that the positions would enhance the officers' careers. In September 1964, the joint NASA–air force committee reported that ninety-four air force officers already worked in NASA and that forty-two of the fifty-five additional positions requested should be filled by further air force assignments. The air force reserved the right to select junior and midgrade officers for NASA tours of at least three years. Eventually, Phillips requested and received assignments for 128 more junior officers, who were mainly assigned to Apollo operations in Houston.[71]

Mueller and Phillips placed officers in key managerial positions throughout the Apollo and Gemini programs, particularly in project control and configuration management offices. Phillips also requested a few senior officers by name. By December 1964, NASA and the air force agreed to assign Phillips as Apollo program director. NASA assigned Brig. Gen. David Jones as deputy associate administrator for Manned Space Flight (Programs), Col. Edmund O'Connor as director of MSFC Industrial Operations, Col. Samuel Yarchin as deputy director of the Saturn V Project Office, and Col. Carroll Bolender as Apollo mission director.[72]

Phillips recognized that NASA's engineers would resist his primary control technique, configuration management. To meet this issue head-on, he scheduled a Configuration Management Workshop in February 1964 in Los Angeles. Each NASA and contractor organization could invite two or three people, except MSFC, which could invite six. At the conference, Phillips stated, "Coming out of Minuteman and into Apollo, I think I've been identified as a procedures and methods man with a management manual all written that I intend to force on the Apollo program and down the throats of the existing 'good people' base. So I ask myself, 'WHY PROCEDURES AND METHODS?'" Phillips noted that good processes and methods made good people's work even better and that they enabled the manager to communicate and control more effectively because they created a "high percentage of automatic action." Because of Apollo's rapid growth, the program needed better communications between NASA and contractor organizations, and all parties had to use consistent language. Phillips stated that good procedures had a high probability of preventing oversights or shortcuts that could lead to catastrophe. He noted, "The outside world is critical, including the contractors, Congress and the GAO [General Accounting Office], and the press."[73] Solid procedures would help to protect NASA from criticism, by preventing failures and by documenting problems as they were found.

Phillips explained his system of design reviews and change control, both of which would help managers control resources. He viewed the field centers as Apollo prime contractors and considered configuration management a contractual mechanism to control industry. Phillips proposed strong project management along with a series of reviews tied to configuration control. Consistent with military doctrine, Phillips believed that "diffused authority and responsibility" meant a "lack of program control," so he assigned responsibility for each task to a single individual and gave that person authority to accomplish the task.[74]

To aid his initiative, Phillips modified Apollo's information system. In the existing headquarters program control and information system, data were collected from the field centers each month, and then headquarters managers reviewed the data for problem areas and inconsistencies, at which time they advised the centers of any problems. Instead, Phillips wanted daily analysis of program schedules and quick data exchange to resolve problems, with data placed in a central control room modeled on Minuteman. He immediately placed contracts for a central control room, which eventually contained data links with automated displays to Apollo field centers.[75]

At MSFC, Saturn managers already had a control room to track official program activities. Saturn V manager Arthur Rudolf assigned names to each chart on the control room wall, so that whenever he found a problem, he could immediately call someone who understood the chart's details and implications. The charts varied over time, as new ones representing current problems and status appeared, replacing charts addressing problems that had been resolved.

This room, although useful, had its problems. Rudolf did not like PERT and reverted instead to "waterfall" charts (Gantt, or "bar," charts arranged in a waterfall fashion over time). Project control personnel translated PERT charts into Rudolf's waterfall charts, located in a small room across the hall. Because the control room displayed the "official" status of items, it did not always have the kind of information Rudolf wanted. A typical problem would be that a hardware item might be completed but the paperwork might still be incomplete. The control room information would not be updated until the paperwork was complete, and hence it did not reflect the true status of the hardware. By contrast, the "mini control room" across the hall had "grease-penciled" charts with more up-to-date information. With MSFC personnel calling contractors to get status updates, this mini control room buzzed with activity.[76]

Phillips devoted great effort to the promotion of configuration management. His headquarters group worked with the air force to develop the *Apollo Configuration Management Manual.* Issued in May 1964, this manual copied the air force's manual, which AFSC officers were updating at the time. He wrote letters to the field center directors emphasizing the manual's importance and directed them to develop implementation plans. By fall 1964, the field centers were actively creating configuration control boards (CCBs), developing the forms and procedures, and directing contractors to implement configuration management.[77]

Configuration management required that NASA have firm requirements and specifications for *Apollo,* which did not yet exist, despite three-year-old contracts between NASA and its contractors. Phillips ordered NASA headquarters and systems contractor Bellcomm to develop definitive specifications. Not wanting Bellcomm to take over this task, field center and contractor personnel quickly became involved, resulting in firm specifications for all Apollo contracts.[78]

NAA recognized that with Phillips as Apollo program manager, failure to enhance configuration management "could well be a serious mistake," so it led a study of the process. Under Phillips's plan, NASA placed preliminary specifications under change control after the Preliminary Design Review, and the hardware under change control after the First Article Configuration Inspection. NAA's group discovered that after the Critical Design Review, NASA did not elevate the final design specifications to contractual status, which could lead to contractual disagreements over design specifications changes. NAA raised this to NASA's attention, resulting in further enhancement of configuration management.[79]

Phillips soon encountered the resistance he expected. Some of it was passive. He tried to educate NASA personnel about systems management through air force project management and systems engineering courses at the Air Force Institute of Technology, where he occasionally lectured. Despite his enthusiasm, only a handful of NASA engineers and managers attended the one-week course.[80]

Other resistance was overt. At the June 1964 Apollo Executives Meeting, Phillips had his headquarters configuration control manager describe configuration management to the field center directors and industry executives. After the presentation, the NASA field center directors argued against Phillips's plan.[81] MSC Director Gilruth had heard that configuration management cost one additional person for every manufacturing team member. If this was true, then configuration management would be far too expensive.

Phillips's team replied that based on Minuteman experience, configuration control required only one person per one hundred manufacturing team members. Air force configuration managers described how Douglas Aircraft took four months to confirm the *S-IV* stage configuration prior to testing and delivery to NASA, whereas on the Gemini Titan II, a program with configuration management, it took two days.[82]

MSFC Director von Braun objected: "This whole thing has a tendency of moving the real decisions up, even from the contractor structure viewpoint, from the one guy who sits on the line to someone else." Boeing President Bill Allen replied, "That's a fundamental of good management." After all, he said, "Who, around this table, makes important decisions without getting advice from the fellow who knows?" Von Braun retorted, "The more you take this into the stratosphere and take the decisions away from the working table—I think the object of this whole thing is to remove it from the drawing board." Allen, whose company had originally created configuration management, replied that you merely had to move the "top engineering guy into the position of the Configuration Manager."[83]

Frustrated in this argument, von Braun retorted that because the military produced a thousand Minutemen, whereas there was only one or just a few Apollos, "We have to retain a little more flexibility." Again, Boeing's Allen disagreed: "Maybe I don't understand, but in my simple mind, it doesn't make any difference with respect to what has been outlined here, whether it's R&D, *Saturn*, or whether you're trying to produce a thousand letters." Von Braun replied, "If you want to roll with the punches, then you have to maintain a certain flexibility." OMSF chief Mueller intervened; on the Titan III program, the first with configuration management from the start, Mueller noted, "Everything is on cost and schedule, even though it married solids and liquids."[84] Mueller concluded, "Configuration management—doesn't mean you can't change it. It doesn't mean you have to define the final configuration in the first instance before you know that the end item is going to work. That isn't what it means. It means you define at each stage of the game what you think the design is going to be within your present ability. The difference is after you describe it, you let everybody know what it is when you change it. That's about all this thing is trying to do."[85]

Mueller, who had the ultimate authority to force implementation, quelled executive objections, but resistance continued in other forms. MSC assigned only one person to configuration management by October 1964, slowing its adoption there and at MSC's contractors. There was continued engineering resistance to change control, Phillips noted: "Engineers always know how to do it better once they've done it, and want to make their product better." Yet "even engineers will admit that changes first of all must be justified," he added.[86]

Contractors had one last chance to charge additional costs to the government before NASA could control them in detail using configuration management. They used the opportunity, preventing full implementation into late 1965 by charging high rates to implement configuration control systems. NASA auditors found numerous deficiencies, including incomplete engineering release systems, no configuration management of major subcontractors, uncontrolled test requirements and procedures, poor numbering systems, lack of documentation, and, in a few cases, no system at all.[87]

With continued exhortation and substantial pressure, Phillips established configuration management on Apollo by the end of 1966, with four different levels of change control and authority: contractor, stage and system manager, program manager, and program office. Contractors could authorize changes that did not affect any other contractors or specifications. Stage and system managers could authorize changes within their own systems. Only program manager offices at the field centers could authorize changes affecting interfaces between stages, systems, or field centers. Phillips in the Program Office authorized changes to the master schedule, hardware quantities, or the overall program specifications. The full system included six formal reviews, after which NASA approved or modified the specifications, designs, and hardware.[88]

Because the Apollo program had been under way for three years before Phillips took control, many of the designs never underwent the initial reviews called for in the new system. NASA had awarded Apollo contracts without accurate specifications, revised the design after contract awards, and designed and even tested system components without specifications. Only one element, the Block II command module for the lunar orbit and landing missions, fol-

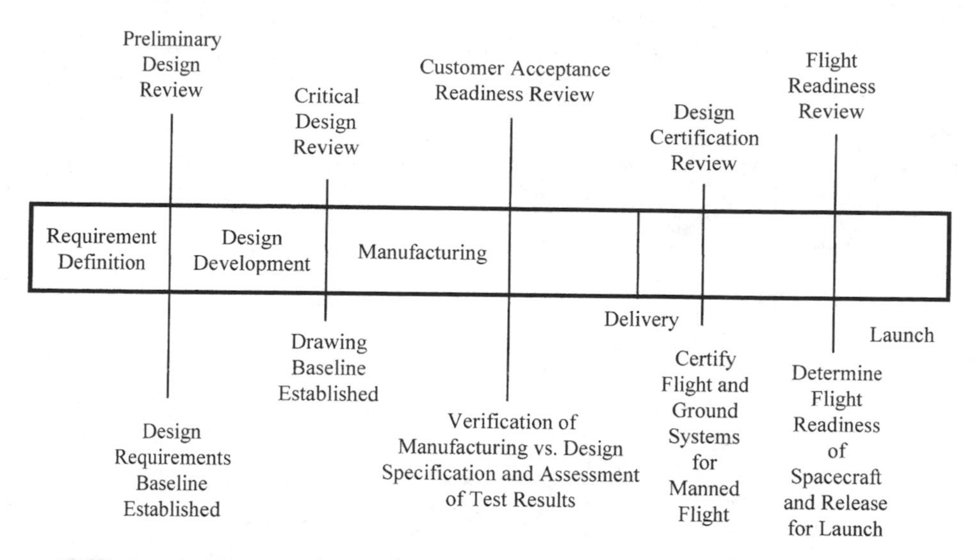

Phillips's review processes for Apollo, which he used to ensure the success of the moon landings. Adapted from Robert C. Seamans Jr. and Frederick I. Ordway, "The Apollo Tradition: An Object Lesson for the Management of Large-Scale Technological Endeavors," in Frank P. Davidson and C. Lawrence Meador, eds., *Macro-Engineering and the Future: A Management Perspective* (Boulder, Colo.: Westview Press, 1982), 20.

lowed Phillips's process in its entirety.[89] Other vehicle elements began their first design reviews at their state of maturity when Phillips levied his requirements.

Phillips augmented configuration control with other information sources. He held daily and monthly meetings with program personnel. In turn, Mueller and Seamans reviewed Apollo each month, and Webb reviewed it annually. The project developed a computer system to automate failure reports, cost data, and parts information. Apollo's Reliability and Quality Assurance organization developed into an important management tool, supplying information on reliability, test results, and part defects as well as current plans, status reports, schedules, funding, and manpower. It forwarded quarterly and weekly highlight reports on each major system element.[90]

The two years following the hiring of George Mueller in September 1963 marked Apollo's transition from a loosely organized research team to a tightly run development organization. Mueller made important early decisions, including instituting his GEM box organization formalizing systems engineering, reliability and quality assurance, and project control on the manned programs. Mueller forced von Braun and Gilruth to adhere to his all-up decision, sharply reduced flight tests in favor of ground testing, and gave more responsibility to contractors. Finally, he started the importation of air force officers to implement program control, beginning with Minuteman director Phillips.

Phillips brought in air force officers to implement configuration management. Configuration management required precise knowledge of the system specifications and design, the baseline against which managers and systems engineers judged changes. This, in turn, required a series of design reviews and managerial checkpoints that progressively elevated specifications and designs to controlled status. Despite resistance, by 1966 Mueller and Phillips augmented NASA's processes by firmly establishing air force methods within OMSF. The new management methods could not prevent all technical problems or make up completely for the earlier lack of management control, nor did contractors uniformly enforce them. For most technical programs, the most difficult times arise when testing uncovers problems. Apollo would be no different.

Smoke, Fire, and Recovery

Apollo's troubles began in September 1965, when NAA's second stage ruptured during a structural test.[91] Engineers pinpointed the fault, and in the process MSFC managers concluded that NAA's management was to blame for shoddy workmanship. By October, the Industrial Operations manager, Brig. Gen. Edmund O'Connor, told von Braun, "The S-II program is out of control." He believed its management was to blame. O'Connor was equally blunt in a letter to Space and Information Systems Division (S&ID) President Harrison Storms: "The continued inability or failure of S&ID to project with any reasonable accuracy their resource requirements, their inability to identify in a timely manner impending problems, and their inability to assess and relate resource requirements and problem areas to schedule impact, can lead me to only one conclusion, that S&ID management does not have control of the Saturn S-II program."[92]

Phillips went immediately to NAA with a "tiger team" of nearly one hundred NASA personnel to "terrorize the contractor,"[93] reporting the team's

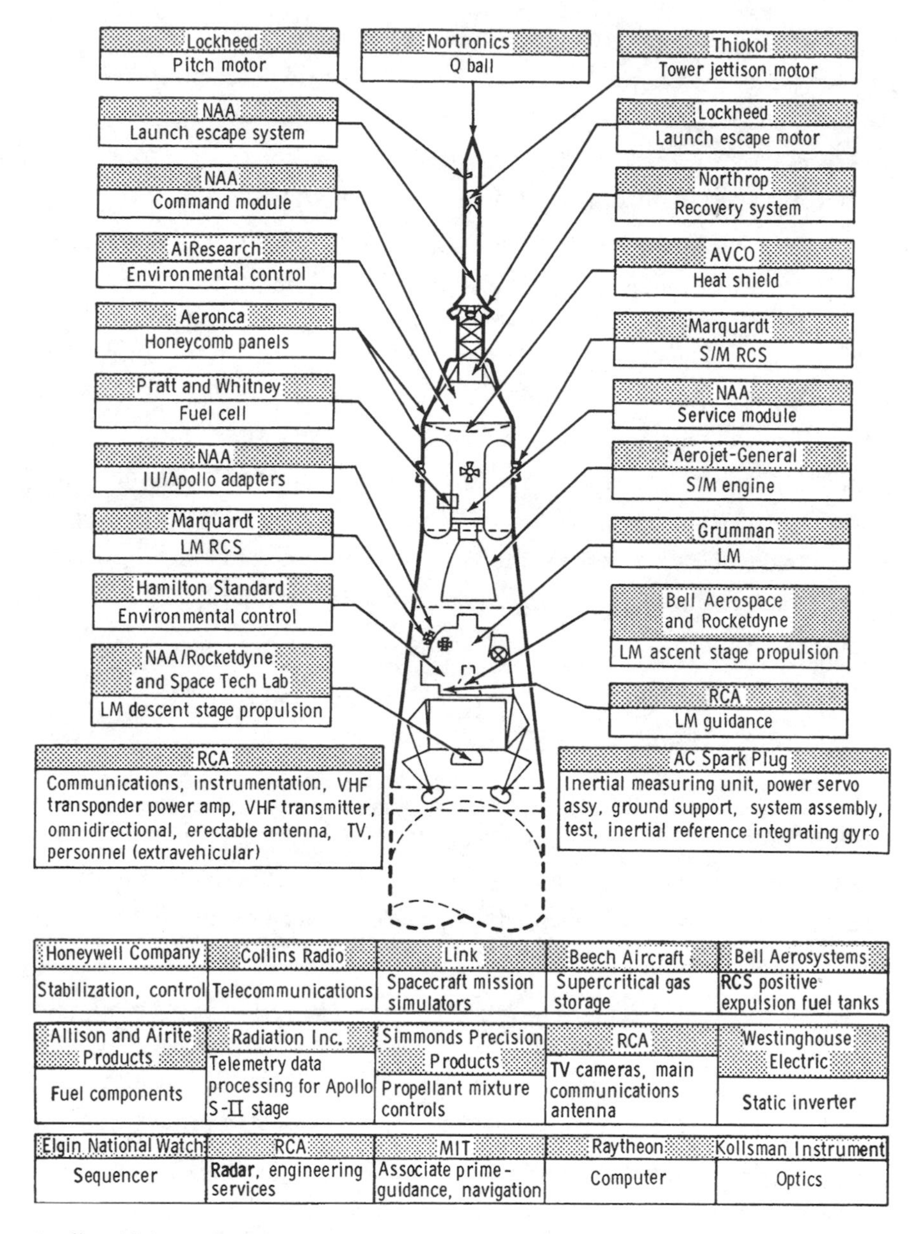

Apollo with its major contractors identified. Apollo was perhaps the largest single R&D project of all time, integrating many contractors for its stages and requiring massive launch and operations facilities and organizations. Saturn V contractors not identified. Courtesy NASA.

findings in December 1965 in what later became known as the Phillips Report. While writing to NAA that "the right actions now" could improve the program, Phillips privately wrote Mueller that NAA's president was too passive. Storms, Phillips said, should "be removed as president of S&ID and be replaced by a man who will be able to quickly provide effective and unquestionable leadership for the organization to bring the division out of trouble."[94]

NAA responded by placing Gen. Robert Greer, retired from the air force, in charge of the S-II program. Greer updated the management control center and ensured more rapid exchange and collection of information through Black Saturday meetings modeled after those in Bernard Schriever's ballistic missile program. Greer also instituted forty-five-minute meetings every morning, eventually cutting back to twice a week. Greer's reforms began to take hold but did not prevent the May 1966 loss of another test stage because of faulty procedures. NASA clamped down further, requiring NAA to develop better methods for managing and planning its work. In the summer of 1966, after two years of studies and preparation, NAA deployed work package management for the S-II and Command and Service Module.[95]

Work package management extended project management to lower project levels and combined accounting and contracting procedures by creating a specific work package for each program task. The company assigned responsibility for each task to one person, a mini project manager for the task who accounted for performance, cost, and schedule in the same way and with the same tools as the overall project manager. Each work package was a "fundamental building stone," with specifications, plans, costs, and schedules to help managers in their monitoring. Prior to the development of work packages, "It was difficult to say what manager was responsible for a particular cost increase because there were 10 or 15 functional and subcontractor areas involved."[96] In later versions, the work package numbering scheme matched that for cost accounting.

Grumman's difficulties on the lunar module also attracted NASA attention. Troubles first appeared in schedule slips on its ground support equipment in the spring of 1966. Alarmed at Grumman's growing costs, Phillips sent a management review team to Grumman that summer, prompting Grumman to sack the program manager, establish a program control office, and move Grumman's vice president to the factory floor to monitor work. By fall, NASA pushed Grumman into adopting work package management.[97] It did not immediately solve Grumman's difficulties. The primary problem was a late start due to NASA's delayed decision to use lunar orbit rendezvous. However, work package management and the new program control office found and resolved problems more quickly than before.

Despite these difficulties, Apollo moved briskly forward until its most severe crisis struck on January 27, 1967. That day, astronauts and KSC personnel were performing tests in preparation for launch of the first manned Apollo mission. At 6:30 that evening, the three astronauts scheduled for that mission, Virgil Grissom, Edward White, and Roger Chaffee, were in the spacecraft command module testing procedures. At 6:31, launch operators heard a cry from the astronauts over the radio, "There is a fire in here!" Those were their last words. All three astronauts died of asphyxiation before launch personnel reached them.[98]

KSC personnel immediately notified NASA headquarters. Administrator Webb hurriedly planned for the political fallout. He sent Seamans and Phillips to Florida, while he persuaded the president and Congress to let NASA perform the investigation.[99] NASA's investigation concluded that the causes of the disaster were faulty wiring, a drastic underestimation of the dangers of an all-oxygen atmosphere, and a capsule design that precluded rapid escape. No one had realized how dangerous the combination was. NASA had used a pure oxygen atmosphere in all of its prior flights, as did air force pilots in their high-altitude flying. As Col. Frank Borman, one of NASA's most experienced astronauts, put it during the Senate investigation, "Sir, I am certain that I can say now the spacecraft was extremely unsafe. I believe what the message I meant to imply was that at the time all the people associated and responsible for testing, flying, building, and piloting the spacecraft truly believed it was safe to undergo the test."[100]

Congress did not prove NASA to be negligent or incompetent. One of the investigation's important results was a nonfinding. Despite searching long and hard, Congress did not find fault with Phillips's management system. Phillips had already uncovered problems with NAA and had been working for some time to make improvements to its organization and performance. The management system used to organize the capsule design was NASA's original committee-based structure, upon which Phillips had superimposed configuration management. He and his management system came out unscathed.

Congressional investigations did uncover some of NASA's dirty laundry, particularly problems with command module contractor NAA. Sen. Walter Mondale of Minnesota learned of the Stage II Phillips Report and confronted Webb about problems between NASA and NAA. Caught by surprise, Webb said he did not know of any such report, which at that moment he did not. After the hearing, he found out about it from Mueller and Phillips. Furious, Webb launched a "paper sweep" to search for more skeletons in the closet. The sweep uncovered a memo written by GE to Apollo spacecraft director Joseph Shea, warning Shea of the danger of fire in the command module. Shea had passed the memo on to his safety and quality assurance people, who responded that no significant dangers existed. GE, already in a sensitive situation because MSC considered it to be spying for headquarters, did not push it any further.[101]

Webb reacted angrily to these revelations. He believed OMSF had been far too independent and secretive. Webb told Seamans, "You have to penetrate the [OMSF] system, don't let Mueller get away with bullshit." The problem, according to Webb, was a lack of supervision by NASA's executive management. Mueller had "followed the policy in Houston of obtaining the very best men they could for the senior positions, and had, as a part of the process of obtaining them, given assurances that they would have almost complete freedom in carrying out their responsibilities."[102]

After Seamans left NASA in late 1967, Webb expressed shock at the poor management system.[103] Webb probably did not realize how decentralized NASA's management really was. Executive managers routinely delegated most decisions to lower levels. In the wake of the fire, this did not seem wise.

When NAA refused to make swift and comprehensive changes—and even expected to be paid a fee for the burned-out spacecraft—Webb called Boeing to see if it would take the job. Boeing said that although it did not want to take over the job, if pressed it would do so. Webb returned to NAA, demanding that it remove S&ID head Storms, further centralize Apollo project management, eliminate any fee for the failed spacecraft, and pay for improvements. NAA did not take the chance that he was bluffing. NAA was extremely unhappy with the entire situation because from its viewpoint, NASA was at fault. Shortly after contract award, over NAA's objections, NASA had directed a change from a nitrogen-oxygen atmosphere to an all-oxygen atmosphere.[104]

One problem uncovered during the investigation was GE's unwillingness to contest NASA over safety issues with a pure oxygen atmosphere. At the heart of the problem was industry's reluctance to confront NASA when industry was dependent on government funding. Despite his substantial political acumen, Webb appeared not to comprehend this. He had hired GE and Bellcomm to strengthen headquarters' ability to monitor the field centers in 1962; after the fire, Webb repeated his mistake by expanding Boeing's role from integrator of the Saturn V to integrator of the entire Apollo-Saturn system to "penetrate the OMSF system." Phillips, who understood the political problems inherent in the GE and Boeing integration efforts, revised the Boeing contract to avoid the negative consequences of Webb's misconception.[105] In essence, Webb wanted to use GE, Bellcomm, and Boeing as an arm of NASA headquarters to control MSC, MSFC, and KSC. This could not work because these contractors could not challenge NASA field center personnel for fear of losing their contracts.

Boeing, as part of its contract, further integrated the management system. The "teleservices network" connected NASA project control rooms with hard copy data transmittal, computer data transmission, and the capability to hold a teleconference involving MSC, MSFC, KSC, Michoud (where the *Saturn I* was manufactured), and Boeing's facility near Seattle. Boeing copied MSFC's program control center design at each facility.[106]

After the fire, NASA placed even more emphasis on achieving high quality and safety through procedural means. In September 1967, NASA set up safety offices at each field center, along with the first project safety plan. The next month, MSC established a Spacecraft Incident Investigation and Reporting Panel to look into anomalies. A month later, NAA created a Problem Assessment Room to report and track problems.

Phillips ordered an astounding array of program reviews to prepare for Apollo's upcoming missions. He wrote to field center managers to ensure that they used the upcoming Design Certification Reviews to evaluate all potential single-point failures.[107] In January 1968, he ordered a complete system safety review, analyzing the interaction of the mission with the hardware, astronauts, ground systems, and personnel. Other reviews included those for quality and metrology, launch vehicle and spacecraft schedules, the communications network, flight readiness, mission planning, subcontractors, site selection, the Lunar Receiving Laboratory, flight evaluations, anomalies, crew safety, interface management, software, and lunar surface activities.[108]

NAA's procedures exemplified the upgraded problem reporting system. Engineers reported failures on a Problem Action Record form. Reliability engineers sent failed components to the appropriate organization, which responded by filling out a Failure Analysis Report describing the physical cause of the failure and the corrective actions taken or recommended. If the organization determined that an engineering change was necessary, it submitted a change request to the change boards. The program control center tracked report status, and a centralized reliability "data bank" recorded the problem and its resolution. Follow-up failure reports and dispositions closed all failure reports.[109]

Another change in the aftermath of the fire was a further strengthening of configuration management, primarily through changing CCB operating procedures. An October 1967 rule disallowed nonmandatory changes for the first command and lunar modules and required the MSC Senior Board to rule on any and all changes to these spacecraft. A February 1968 ruling required managers to consider software changes and their ramifications in CCBs. In May 1968, Apollo Spacecraft Manager George Low specified that the MSC CCB had authority over all design and manufacturing processes.[110]

By 1968, tough CCB rules slowed the program as trivial changes came to the attention of top managers. Eventually, even Phillips realized that centralization through configuration management could go too far. In September, MSC managers classified changes into two categories: Class I changes, which MSC would pass judgment upon, and Class II changes, which could be approved by the contractors. Classification did not by itself help much, so in October 1968 Phillips gave Level II CCBs more authority, while higher levels ruled on schedule changes.[111]

The Apollo project met its technical and schedule objectives, landing men on the Moon in July 1969 and returning them safely to Earth. Anchored by configuration management, Phillips's system weathered the storm of problems uncovered through testing and Apollo's most severe crisis, the 1967 death of the three astronauts and the ensuing investigations. Despite strenuous efforts, congressional critics did not find many flaws with Phillips's management scheme and concurred with NASA that the fire resulted from a tragic underestimation of the danger.

Configuration management was Phillips's most powerful tool. Whenever problems occurred, his almost invariable response was to strengthen configuration management. Having found that his favorite method could be overused, by the end of 1968, Phillips gave lower-level CCBs more authority. Configuration management formed the heart of Apollo's system and has remained at the core of NASA's organization ever since.

Von Braun's Conversion

Despite the imposition of systems engineering and configuration management on MSFC, they remained foreign concepts to members of Huntsville's engineering family. They had been working together on rockets since the 1930s, and the many years of experience had taught them the technologies, processes, and interactions necessary to build rockets. These engineers understood rocketry and each other so well as to make formal coordination mechanisms such as systems engineering redundant. The efforts of Mueller and Phillips had brought configuration management and air force methods to contractors, but the functional, discipline-based laboratories remained the centers of power in MSFC.

As MSFC's effort on the Apollo program peaked in 1966 and layoffs threatened, however, MSFC leaders realized that they would have to diversify beyond rocketry to keep themselves in business.[112] In new fields such as manned space stations and robotic spacecraft, MSFC's unmatched ability in rockets meant little, and they soon found new utility in systems engineering.

By the summer of 1968, von Braun recognized that he needed to strengthen systems engineering at MSFC. He called in Philip Tompkins, a communications expert from Wayne State University, to study MSFC's organization and recommend how better to implement systems engineering. At the time,

Mueller was pressing von Braun to emphasize systems engineering in the design of the *Skylab* space station. Von Braun explained to Tompkins that Mueller, who had been trained in electrical engineering, thought more naturally in terms of a "nervous system" than he, who thought of rockets as machines. Von Braun belatedly saw the validity of Mueller's point of view and was determined to reorient MSFC along systems engineering lines so as to better coordinate MSFC's design efforts.[113]

Tompkins investigated MSFC's organization and soon concluded that the design laboratories were overly oriented toward "low-level subsystems engineering." As one manager stated it, "If we had a lawnmower capability at the Marshall Center, we'd put lawnmowers on all the vehicles."[114] To combat this, Tompkins recommended significant strengthening of the systems engineering office. With this change, systems engineering by late 1968 became a much stronger element within MSFC, albeit weaker in the traditional rocket groups than the newer organizations that focused on other projects. As MSFC's "family" organization and expertise in rocketry grew less important, systems engineering took their place. Formal coordination processes replaced the informal methods that sufficed in von Braun's heyday.

Why did NASA's most experienced group of engineers take so long to embrace systems engineering? Three factors contributed: the almost exclusive use of in-house capability for rocket development and testing, the extraordinary continuity of von Braun's team, and the continuity of the team's R&D project. From the mid-1930s until the early 1960s, von Braun's team members relied upon their own capabilities to design rockets, using external contractors sparingly. When they did use contractors, they did so for only specific components, or they closely monitored the contractors, such as Chrysler for the Redstone and Jupiter. The use of contractors significantly increased for the Saturn V project, and systems engineering began to make inroads into MSFC at this time. However, not until MSFC diversified out of rocketry and many of the original team began to retire in the late 1960s did systems engineering become a major element of Marshall's R&D process.

The continuity of von Braun's team, along with the continuity of the technologies upon which the team worked, helps to explain the dismissal of systems engineering at MSFC. Simply put, when each team member knew the job through decades of experience and knew every other team member over that period, formal methods to communicate or coordinate were redundant. Rocket team members knew their jobs, and each other, intimately. They understood what information their colleagues needed, and when. When they began to work on new products such as space stations and spacecraft in the late 1960s, it was no longer obvious how each team member should communicate with everyone else. Formal task planning, coordination, and communication became a necessity, and systems engineers performed these new tasks.

Conclusion

Systems management evolved as the manned space programs developed. Like the ballistic missile programs before them, the manned programs were inaugurated with few cost constraints and substantial external pressure to speed development. Despite massive cost overruns, the programs continued for the first few years with few questions from headquarters or Congress. Glennan, and later Webb, let the STG, MSC, and MSFC do their jobs with minimal supervision. These organizations used informal engineering committees to manage the manned programs. When NASA needed rigor in manufacturing and component quality, it had the air force and its industrial contractors to supply them. Informal methods frequently produced technical success but failed miserably at predicting costs.

Spiraling costs led Holmes, the first head of OMSF, to challenge Webb. Holmes's failure made it obvious to his replacement, Mueller, that he had to control costs. To do so he enlisted the help of air force officers, led by Phillips. Mueller forced MSC and MSFC to adopt stronger project management, institute systems engineering, expand ground testing, and report more thoroughly to headquarters. Phillips instituted configuration management and project reviews throughout Apollo to control technical, financial, and contractual aspects as well as the scheduling of the program. Air force officers brought in by Mueller and Phillips propagated the reforms and transformed OMSF's organizations into project-oriented hierarchical development organizations.

Systems management made development costs more predictable and created technically reliable product, but at a price. The disadvantages of systems management would become apparent later, but for the moment it was a managerial icon. If there was a secret to Apollo, it was Phillips's organizational reforms, which transferred air force methods to NASA, superimposed upon the technical excellence of STG and MSFC engineers. Europeans would eventually make a concerted effort to learn the managerial secrets of Apollo, but not before trying their own ideas, and failing miserably.

SIX

Organizing ELDO for Failure

> The failure of F11 in November 1971 brought home to the member states—and this was indeed the only positive point it achieved—the necessity for a complete overhaul of the programme management methods.
>
> —General Robert Aubinière, 1974

World War II left Europe devastated and exhausted, while the United States emerged as the world's most powerful nation, both militarily and economically. Western Europeans feared the Soviet Union's military power and totalitarian government, but they worried almost as much about America's immense economic strength. Some asserted that American dominance flowed from the large size of American domestic markets or the competitive nature of American capitalism, while others believed that technological expertise was the primary force creating "gaps" between the United States and Europe. By the late 1960s, the "technology gap" was a hot topic for politicians and economists on both sides of the Atlantic.

Investigations showed that European technology and expertise did not radically differ from that of the United States. However, a number of studies showed that Americans managed and marketed technologies more efficiently and rapidly than Europeans. Significant differences between the United States and Western Europe existed in the availability of college-level management education and in the percentage of research and development expenditures. In each of these areas, Americans invested more, in both absolute and per capita terms. Some analysts believed the technology gap to be illusory but a management gap to be real.

To close the gaps, Europeans, actively aided by the United States, took a number of measures to increase the size of their markets, to develop advanced science and technology, and to improve European management. The Common Market was the best-known example of market integration. Science and technology initiatives included the Conseil Européen pour Recherche Nucléaire (CERN [European Committee for Nuclear Research]) for high-energy physics research, EURATOM for nuclear power technologies and resources, the European Space Research Organisation (ESRO) to develop scientific satellites, and the European Space Vehicle Launcher Development Organisation (ELDO) to create a European space launch vehicle.

Because of the military and economic significance of space launchers, the national governments of the "big four" Western European states—the United Kingdom, France, West Germany, and Italy—all supported the European launcher effort. Seeking contracts, the European aircraft industry also actively promoted the venture. Paradoxically, these strong national interests rendered ELDO ineffective. Each country and company sought its own economic advantages through ELDO, while withholding as much information as possible. This attitude led to a weak organization that ultimately failed. When the Europeans decided to start again in the early 1970s, ELDO's failure was the spur to do better, a prime example of how not to organize technology development.

The American Challenge

The European fear of gaps between themselves and the superpowers derived from changed political, military, and economic realities after World War II. Germany was devastated, occupied, and dismembered. Italy was torn between its Fascist past, the resistance movement led by the Communists, and the Catholic Church. France had been defeated by Nazi Germany, then riven by hostilities between the Vichy regime, Charles de Gaulle's Free French, and the Communist Party. Britain was victorious, but the war depleted its treasury and exhausted its people. By contrast, the Soviet Union's armies advanced into and remained in Central Europe. The United States emerged from the war with sole possession of the atomic bomb, a booming economy, and growing resources. The Soviet Union and the United States became the superpowers, relegating Western Europe to second-tier status.

Many American diplomats believed a strong, united Europe was the best means to defend against Soviet military expansion or internal chaos that might lead to Communist takeover. To strengthen the German economy without antagonizing France, American diplomats in 1947 offered the Marshall Plan to European countries on the condition that they work together to allocate funds. This led to the creation of the Organization for European Economic Cooperation, which later became the Organization for Economic Cooperation and Development (OECD). The Communist coup in Czechoslovakia and the Berlin blockade inaugurated negotiations that led to military cooperation with the North Atlantic Treaty Organization in 1949.[1]

European leaders also sought to cooperate with each other, apart from the United States. France decided to control German ambitions by forming a strong alliance with its historic enemy. The Low Countries, which needed a strong European economy with which to trade, and the Italians, who needed an outlet for unemployed workers and access to technology and natural resources, combined with France and West Germany to form the European Coal and Steel Community in 1950. The same countries agreed to the Common Market in 1957, which lowered mutual tariff barriers and created the large market they believed critical for economic growth.

Nuclear technology development also benefited from European integration. Physicists Pierre Auger of France and Edoardo Amaldi of Italy led efforts to create a European laboratory for high-energy physics research to compete with U.S. physics researchers. European research leaders agreed to create CERN in February 1952 to develop a large particle accelerator and supporting facilities. The need to distribute and control uranium for nuclear reactors led to the creation of EURATOM in the 1957 Treaty of Rome that created the Common Market. Despite these European efforts to enlarge their market and pool resources for nuclear technology, American developments in electronics and computers, and the American and Soviet development of rocketry and missiles, appeared to keep the superpowers several steps ahead.[2]

In 1964, French journalist Jean-Jacques Servan-Schreiber wrote a book that served as a manifesto to European governments and industry: *The American Challenge.* He described the penetration of American industry into Europe and argued that the cause was not "a question of money." Rather, the United States had developed better and more widespread education, leading to more flexible policies and management. As he put it, "Europe's lag seems to concern *methods of organization* above all. The Americans know how to work in our countries better than we do ourselves. This is not a matter of 'brain power' in the traditional sense of the term, but of organization, education, and training."[3] Significantly, he illustrated American dominance by using the examples of the computer and aerospace industries.

Servan-Schreiber had influential readers on both sides of the Atlantic. U.S. Secretary of Defense Robert McNamara thought "the technological gap was misnamed," believing it to be a managerial gap. Europeans needed to develop their educational systems. McNamara noted, "Modern managerial education —the level of competence, say, of the Harvard Business School—is practically unknown in industrialized Europe."[4] West Germany's defense minister, Franz Josef Strauss, also agreed with the French journalist, believing the technology gap was due to advances in space technology, computers, and aircraft construction, three areas he thought decisive for the future. Because large corporations performed the majority of research and because of the large domestic market, American companies had the advantage of scale over their European competitors. Strauss's solution was to create an integrated European community with common laws and regulations and to pool European resources. European countries needed to launch large, multinational high-technology projects to provide opportunities for European corporations to work on big research and development endeavors.[5]

Both Servan-Schreiber and McNamara believed that their societies needed more and better management. According to McNamara, "Some critics today worry that our democratic, free societies are becoming overmanaged. I would argue that the opposite is true. As paradoxical as it may sound, the real threat to democracy comes not from overmanagement, but from undermanagement. To undermanage reality is not to keep it free. It is simply to let some force other than reason shape reality."[6]

Servan-Schreiber's views were similar: "*Only a deliberate policy of reinforcing our strong points*—what demagogues condemn under the vague term of 'monopolies'—*will allow us to escape relative underdevelopment.*" But he did make one allowance: "This strategy will rightly seem debatable to those who mistrust the influence and political power of big business. This fear is justified. But the remedy lies in the power of government, not in the weakening of industry."[7] Given the economic and military importance of technology development, Servan-Schreiber and others accepted the risks of government and business power.

Academic analysts of the technology gap used more sophisticated means to reach similar conclusions. Analyses by the OECD generally recommended increases in the number and scale of integrated European high-technology projects. The technology gap "is not so much the result of differences in technological prowess, except in some special research-intensive sectors," said economist Antonie Knoppers, "as of differences in management and marketing approaches and—possibly above all—in attitudes."[8] He believed the disparities were in "middle or lower levels" of management. Economist Daniel Spencer believed that a "more fruitful way of assessing the technological gap" was "to define it as a management gap" because American managers were "alert to opportunities created by the research of a military or nuclear or space type."[9]

Officials debated about the existence of the gap, its causes, and solutions. British leaders worried about an American "technological empire" and a "brain drain" of technical experts to the United States. The French feared the loss of economic and cultural independence. German leaders worried that the gap exposed flaws in German education and management. American politicians minimized the significance or existence of a technological gap, shifting the argument to differences in culture and management. In 1967, President Lyndon B. Johnson sent Science Advisor Donald Hornig to Europe with a team of experts to study the problem and asked the National Aeronautics and Space Administration (NASA) to explore new cooperative ventures with the Europeans to ease their fears. All agreed that European countries needed educational reforms to bridge the various gaps. This would take a while to accomplish, so in the meantime, the most obvious idea was to mimic American management methods on large-scale technology programs.[10]

In the early 1960s, space launchers beckoned as a particularly fruitful field for integrated efforts. Developing space launchers would eliminate European dependence on the United States to launch spacecraft and also aid national efforts to develop ballistic missiles. A multinational launcher program would teach European companies to manage large programs and spur Europe's educational system to become more responsive to advanced technology and management. Many advantages would accrue, if Europeans could overcome their differences.

European Rocketry and the Creation of ELDO

Development of rockets began before World War II in a number of European countries. Most important was the German program leading to the *A-4* ballistic missile, better known as the *V-2.* Like other early rocket projects, it originated with amateurs in the late 1920s. Just prior to Hitler's ascension to power in January 1933, Army Ordnance coaxed young engineering student Wernher von Braun to be the technical director of its rocketry program. Army Ordnance drew a veil of secrecy over the project, and the Nazi regime soon began to pour large sums into it. Von Braun's team successfully developed the *A-4* rocket, which terrorized the populations of Antwerp and London in 1944 and 1945. Although its military impact was limited, it caught the attention of military technologists around the world.[11]

After the war, each Allied power acquired German rocket technology and experts. The United States acquired the lion's share of both, with rocket parts for more than sixty *A-4*s and program leaders including von Braun and Arthur Rudolf. The Soviet Union acquired a large number of technicians, a few of the leaders, and parts for some twenty *A-4*s. With German assistance, Britain launched three *A-4*s in October 1945 from a site near Cuxhaven. The French acquired a small group of experienced Germans from Peenemünde, who began working at Vernon on an *A-4* derivative vehicle and a new rocket engine.[12]

In March 1949, the French Directorate for Armament Studies and Fabrication decided to build *Véronique,* a single-stage, liquid-fueled sounding rocket. From 1951 to 1964, French engineers extended the design, improving altitude performance from 2 to 315 kilometers. They initially tested this unguided rocket in southern France, later testing it at Hammaguir in the Algerian desert under the direction of Col. Robert Aubinière.[13]

After the 1956 decision to build an indigenous nuclear force, missile development expanded rapidly. French engineers began development of larger rockets capable of placing a small satellite in orbit. In 1960 the French state rocket consortium, Société pour l'Étude et la Réalisation d'Engins Balistiques

(SEREB [Society for Study and Development of Ballistic Engines]), concluded that it was possible to build such a rocket, eventually known as the *Diamant*. The French created their own space agency to fund the launcher project, while SEREB developed the stages and the military tested them. With some assistance from Col. Edward Hall, the U.S. Air Force officer who initially developed the *Minuteman* intercontinental ballistic missile, these efforts came to fruition with France's launch of a small test satellite in November 1965 from Hammaguir, making it the third country to launch a satellite.[14]

The British were also active in rocket design and in the development of nuclear weapons to place on rockets. They tested their first fission bomb in 1952 off the coast of Australia and their first fusion weapon in May 1957 over Christmas Island. From the late 1940s on, they developed a number of missiles, including air-to-surface, surface-to-air, air-to-air, and ship-to-air weapons. When in 1954, U.S. Secretary of Defense Charles Wilson offered to collaborate with the British on a ballistic missile, the British expressed interest and began their own studies. The Americans allowed the formation of agreements between the British company DeHavillands and Convair on the missile structure, and between Rolls Royce and North American Aviation for the engines. Based on these agreements, the British developed a large liquid-fueled rocket known as *Blue Streak*.[15]

It soon became apparent to the British, as it had to the Americans, that liquid-fueled rockets were poor weapons because of their immobility, cumbersome logistics, and long preparation time to launch, which made them vulnerable to a Soviet first strike. American missile efforts quickly surpassed those of the British, and in 1956 the United States offered to place *Thor* missiles in Britain five years sooner than *Blue Streak* would be available. British officials accepted the offer in 1958, and to avoid duplicating *Thor*'s capabilities, increased *Blue Streak*'s required range to 2,500 miles. In April 1960, British military leaders canceled *Blue Streak* in favor of purchasing American air-launched *Skybolt* missiles and sea-launched *Polaris* missiles.[16]

Blue Streak's cancellation as a weapon led British officials to consider its potential as a satellite launcher. Technically this was feasible. The key questions were political and economic. First, the British had sunk £60 million into the project, which needed £240 million more. Such large expenditures would divert scarce funds and technologists from other scientific and technological endeavors. Second, the technology could be obsolete by the time it was completed. On the other hand, Britain would no longer depend on the United States to launch satellites, an important advantage if communication satellites became commercially viable. Prime Minister Harold Macmillan wanted to use *Blue Streak* to forge closer relationships with France. Needing French support to join the Common Market, Macmillan calculated that a joint launcher program with France would smooth Britain's application process. Supported by the American leaders, who continued to favor European integration, he decided to approach the French in late 1960.[17]

The French reaction was cautiously optimistic. French military leaders expressed keen interest in gaining access to inertial guidance technology and nose cone reentry technologies. Because the United States prohibited the export of these technologies to France, this could have been a fatal objection. In other words, if the French insisted on acquiring inertial guidance and reentry technologies as part of the deal, there would be no deal. Unexpectedly, French President Charles de Gaulle threw himself behind the project, even without the military technologies. De Gaulle saw the project as a means to fulfill French technological ambitions, using space and nuclear programs to create a permanent "technological revolution" to support a strong and independent France. Because the project supported this goal, he supported the project. In a meeting with Macmillan in January 1961, he agreed to join the project and to jointly approach other European governments, under the condition that the launcher's second stage be French. Macmillan and de Gaulle scheduled a conference the next month in Strasbourg to broach the subject with other governments.[18]

The Germans accepted an offer to build the launcher's third stage. This gave them the opening to rocketry that they would not undertake alone because of the Nazi *V-2* heritage. That left the question of the Italians, who were already building sounding rockets under American license and had just begun their own launch program with a sea-based launching platform in collaboration with the United States. The British, who were by fall 1961 desperate for an agreement because of their financial problems, put substantial diplomatic pressure on Italy to join. Negotiations produced the convention for ELDO in March 1962, with Italy to build the satellite test vehicle.

Britain paid a heavy price for its desperation, stuck with 38.79% of contributions to the £70 million Initial Programme, scheduled for completion by the end of 1965. France, Germany, and Italy paid 23.93%, 18.92%, and 9.78%, respectively, and Belgium and the Netherlands paid 2.85% and 2.64% to build the ground and telemetry equipment. The convention came into force in February 1964 after Britain, France, and West Germany ratified it.[19]

ELDO's structure emphasized national interests. Because ELDO based contributions on existing programs, Britain and France insisted upon managing their stages through their national government organizations, according to their own procedures. ELDO provided but did not control the funding. Because member states contributed funding in fixed proportions but spent it according to costs, each country had a built-in incentive to increase costs to recoup its investment. For example, if Belgium overran its budget by 50%, it would contribute only its 2.85% share to that overrun.[20] Member states severely circumscribed ELDO's authority, rendering cooperation difficult at best. The job of the Secretariat required delicate negotiation skills, a fact recognized in the appointment of Italian ambassador R. Carrobio di Carrobio to the post. When ELDO came into official existence in February 1964, Carrobio would need all of his diplomatic talents.

Aging Technology and Changing Objectives

Without any staff members except for those supplied by the national governments, and with technical problems more complex than originally thought, ELDO's Preparatory Group moved the project, now known as Europa I, slowly forward between 1961 and 1964. Although the first and second *Blue Streak* launches in 1964 succeeded, delays and technical difficulties led to substantial cost escalation, from the 198 Million Accounting Units (MAU)[21] originally budgeted (the equivalent of £70 million) to 400 MAU at the end of 1964.[22]

Commercial communication satellites soon troubled ELDO executives and national representatives. Americans tested their commercial viability with the *Early Bird* satellite in 1965 and controlled the market through Intelsat, the international consortium for satellite telecommunications. Europeans wanted to break the American stranglehold. In 1964, the ELDO Secretariat reported that an upgraded *Europa I* launcher could place 20–40 kilograms of equipment in polar orbit and that a more powerful *ELDO B* rocket could place a 1,000-kilogram communications satellite in geostationary orbit in the 1970s. The next year, French officials proposed that ELDO scrap *Europa I* and instead immediately begin work on *ELDO B*. Other delegates vetoed this as too risky but agreed to reconsider it later.[23]

Spurred by massive overruns that they disproportionately funded, the British reversed their strong support of ELDO in April 1966. They now believed that *Europa I* would be obsolete and its commercial potential limited and stated that they would neither participate in rocket upgrades nor contribute beyond existing commitments. Under pressure from the other delegations, the British agreed in June 1966 to remain in ELDO, but only if the organization reduced the British contribution. In July, delegates agreed to this but also voted to fund an equatorial base, inertial guidance improvements, and the *Europa II* rocket, which could launch up to 150 kilograms into geostationary orbit. The new cost was 626 MAU, more than three times initial ELDO estimates. Britain would not contribute to any costs above the agreed 626 MAU level.

Technical problems soon pushed costs past the limit. In February 1968, after two failures of the French second stage, the British invoked ELDO's new procedures for projected cost overruns. Two months later, the British announced that they would make no further contributions to ELDO. Italian delegates, angered by the refusal of France and Germany to include them in the bilateral *Symphonie* communications satellite, and also by their inability to recoup ELDO contracts for their own space industry, refused to agree to a French-German "austerity plan" that would have cut Italian portions of the program. After yet another rocket failure in November 1968, this time of the German third stage, delegates from France, Germany, Belgium, and the Netherlands agreed to make up the shortfall in British and Italian contributions to complete a scaled-back program. Italy finally agreed to rejoin, but Britain would supply its first stage for only two more test flights. After that Britain would be through with ELDO. The remaining partners agreed to fund studies for a first stage replacement.[24]

NASA International Programs chief Arnold Frutkin noted the "half-hearted and mutually-suspicious character of participation by its [ELDO's] members." European governments held together only insofar as the United States resisted European commercial interests. With Europeans united in their suspicions that the Americans intended to monopolize communication satellites, Frutkin believed, "US offers in space and other fields of technology will continue to be regarded with extreme and often irrational suspicion until the comsat issue is resolved." By sending mixed signals, American leaders helped keep ELDO alive.[25]

Changing technological objectives contributed to ELDO's problems. Without firm objectives at the start, ELDO and French studies showed that ELDO needed a more powerful launch vehicle to place communications satellites into orbit. The French, who viewed ELDO as an essential part of their drive for independence from the United States, were willing to pay the price. So too were the Germans, who subsidized their own reentrance into rocketry. The Belgians and Dutch believed they needed to go along with their powerful neighbors. As long as ELDO guaranteed the Italians technically interesting tasks, Italian leaders would contribute. However, the British had little to gain in that the first stage was operational. Convinced of American willingness to launch European satellites, British leaders believed it more important to fund applications satellites than launchers. These differences might have been overcome if ELDO's rockets had proved successful.

Organizing for Failure

As was typical for other large projects in Europe and the United States, ELDO managers distributed tasks to a number of organizations. ELDO funded the British Ministry of Aviation for the first stage. The ministry, in turn, channeled funds to the Royal Aircraft Establishment, which contracted with De-Havillands for most of the stage and with Rolls Royce for the engines. De-Havillands subcontracted with Sperry Gyroscope for the guidance package. ELDO funded the French Space Agency for the *Coralie* second stage. The agency, in turn, contracted with the French Army Laboratory for Ballistic and Aerodynamic Research for second stage development and the government's National Company for Study and Construction of Aviation Engines for engines.[26]

West Germany, which had no space organization prior to the ELDO discussions, initially placed its space activities under the German Ministry for

Atomic and Water Power. In August 1962 the Germans formed the government-owned Space Research Company to study space activities, under the guidance of the German Commission for Space Research. The Ministry for Atomic and Water Power expanded to become the Ministry for Scientific Research, under the Ministry for Education and Science. ELDO delegated the Ministry for Education and Science as the national agency to oversee the *Europa I* third stage, and this agency in turn contracted with the newly created industrial consortium Arbeitsgemeinschaft Satelitenträger (ASAT). ASAT was an uneasy—and according to some, involuntary—alliance between two major German aerospace firms, Messerschmitt-Bölkow-Blohm (MBB) and Entwicklungsring Nord Raumfahrttechnik (ERNO). Similar unwieldy arrangements held for the Italian, Belgian, and Dutch portions of ELDO.[27]

Complexity of the organizational structure contributed to ELDO's difficulties but was not the most significant problem. The difficulty was that working groups, usually private companies and government-industry consortia, did not report to ELDO but to their national governments. These, in turn, reported to the ELDO Secretariat in Paris. Because ELDO distributed funds to the national governments, which distributed them using their own procedures, the industrial and engineering groups took their orders from the national governments, not the ELDO Secretariat. This "indirect contracting" structure interposed an extra layer of bureaucracy and gave that layer final authority.[28]

The Secretariat had no authority to force governments or contractors to make changes; it could only make suggestions to the national governments. Nor could the Secretariat take legal action, both because contractual authority lay with the national governments and because the contracts themselves were vaguely worded. As late as 1972, the Europa II Project Review Commission stated, "There is no clear definition of responsibilities within the ELDO organisation, nor between ELDO staff and ELDO contractors."[29]

Uncertainty about roles and responsibilities led to two kinds of situations. When the national agency was "strongly structured," as in Britain and France, it led to "a complete effacement of the Secretariat's role." On the other hand, when the national agency was weak, as in the case of Germany's new organizations, it led to "confusion in the minds of firms about the technical responsibilities of the Secretariat and those of the national agency." In some cases the Secretariat "did not respect the responsibilities of the national agencies" and undermined their authority.[30]

Having unclear and changing requirements did not help. The 1972 review commission concluded, "*Europa II* seems in a continuous state of research and development with major changes made from one launch to the next almost independently of whether the previous flight objectives have been achieved." No single, complete specification existed for the entire vehicle. Without clear specifications, engineers did not have clear goals for defining telemetry measurements, for limiting the weight of the vehicle, or for ensuring quality and redundancy across the project. The end result was "a launch vehicle with little design coherence, and posing complicated integration and operational problems."[31]

Because the British and French designed their first and second stages before ELDO existed, they ensured that their government organizations determined methods and standards for their own stages. ELDO itself had no authority to impose standards. This led to inconsistent and incomplete specifications, documentation, quality standards, and procedures. The Secretariat had no quality organization until 1970, relying upon national teams to enforce good manufacturing practices, use high-quality components, and adhere to testing procedures. At best, the result was components, processes, and documentation of variable quality. In practice, variable quality led to flight failures.[32]

With only a small engineering staff, the Secretariat's ability to analyze problems was also limited. Before ELDO came into official existence, the Preparatory Group relied on engineers supplied by the national governments. After February 1964, the Secretariat built a small engineering staff in the Technical Directorate. Often the engineering staff "endeavored to promote the solution of technical problems, but in some cases important solutions [were] refused on budgetary grounds." Without access to necessary information, adequate staff, or authority to make changes, the Technical Directorate performed little systems engineering. Unless contractors resolved interface problems among themselves, the problems remained unresolved.[33]

Problems lingered in this way because of poor communication. No single location existed for project documentation. Nor did ELDO define what documentation should be produced. Project reviewers noted that "while certain documents were available, there was nothing systematic about this."[34] For communication across national boundaries, barriers of language, industrial competition, and national factionalism took precedence. The most extreme case was with the German third stage contractor ASAT, which had "total disinterest in the IGS [Inertial Guidance System—built by British contractor Marconi], a refusal to attend acceptance or bench integration tests, a lack of cooperation in defining strict working procedures, a total refusal of responsibilities." The ELDO Secretariat failed to bridge the gap between ASAT and Marconi. Communications between manufacturing and testing were poor, as were communications between the launch and engineering teams. In the case of the guidance systems, Marconi "built a wall between users and manufacturers, a wall which was accepted, if not liked, by everybody and which ELDO, among others, did not make much effort to destroy."[35]

The ELDO Secretariat's financial and scheduling groups were better staffed than its technical teams, but the problems were similar. ELDO created a Project Management Directorate, which used tools such as the Program Evaluation and Review Technique (PERT) to track three levels of schedules: the contractors, national programs, and the ELDO Secretariat.[36] Unfortunately, the Secretariat had no authority to force timely or accurate reporting. Analysts lamented, "The reports of the member states are always late."[37] Even when the Secretariat could acquire timely data, it could do little more than watch the schedule slip and remind offenders that they were deviating from the plan. Tools and organizations to report schedule slips and cost overruns were of little use to personnel in the Secretariat, other than to remind them of their lack of power with respect to the national governments and contractors.

By design, ELDO's member states created a weak organization. ELDO's Secretariat had few staff members and little authority to do anything but watch events happen and try to coordinate its unruly member states and contractors. When troubles came—and come they did—the Secretariat tried to coordinate and plan around the problems. What it could not do was manage or control them.

"Paris, We Have a Problem"—with Interfaces

ELDO's technical troubles traced in most cases to problems with interfaces: component boundaries that were also organizational boundaries. Here ELDO's inability to either impose standards or ensure communication among its engineering groups produced its logical result: failure. ELDO engineers and managers soon recognized that they had a major problem with interfaces and communications. Through the Secretariat and national organizations they tried to make this point to the politicians who governed ELDO. Despite efforts to improve ELDO's communications and systems engineering, ELDO's basic flaw was a lack of authority that no piecemeal measures could repair. Symptoms of this problem became evident first in cost overruns and schedule slips, then in flight failures.

In 1964 and 1965, the first test flights of Britain's *Blue Streak* were deceptively promising.[38] DeHavillands's first stage design incorporated several years of design experience prior to the formation of ELDO, as well as American techniques from the Atlas program, which itself had developed for a number of years before Britain acquired some of its technologies. Because the first tests flew only the British first stage, they did not involve interfaces with any other stage. Because a single government organization with prior rocket and missile experience managed *Blue Streak,* and because firms experienced with these technologies and with each other built it, communications were not a problem. *Blue Streak*'s success was not to be repeated.

Problems soon appeared in the interfaces between the rocket stages and the organizations responsible for them. Under ELDO agreements of 1963, member states divided interface responsibility by having the lower stage contractor responsible for interfaces between any two stages. Thus the British were responsible for the interface between the first and second stages, the French for the interface between the second and third stages, and the Germans for the third stage–test satellite interface. Meetings in 1965 further defined interface procedures, specifying that the Interface Design Authority (the lower stage contractor) would freeze the design, make the information available to all parties, and provide for hardware inspection. The Interface Design Authority would submit a Certificate of Design to ELDO to certify the correctness of the interface. However, the scheme had a fatal flaw: "It was not the intention that an Interface Authority should do again work already allocated and being performed by another Design Authority. The Interface Design Authority would therefore base his design declaration on statements, made by the other Design Authorities concerned, that the relevant specifications had been met."[39]

The contractor documenting the interface merely ensured that the other organizations involved provided the appropriate documentation, but no organization analyzed both sides of the interface for discrepancies. ELDO documented the interface specifications and trusted contractors on each side of the interface to abide by them. Without anyone checking *both* sides, misunderstandings about specifications went unnoticed until the organizations tried to connect the stages or test them in flight.

Misunderstandings became painfully evident the moment contractors tried to connect hardware. In an early test of the interface between the French second stage and the German third stage, the structure failed because of the wrong kinds of connecting bolts. When the ELDO Secretariat decided to make changes to the French second stage, the French complained, questioning who had the "power to impose a solution." In another case, the Germans developed a table to mimic the structural interface of the Italian test satellite "before the Italian Authorities had completed their examination [of] the requirements." This led to a mismatch between the assumed size of the connecting ring and the actual ring later designed by the Italians, and the Germans had to scrap their hardware and build a new table.[40]

Complaints about communications and integration problems reached the ELDO Council through member state delegations, leading to a study of ELDO's organization in early 1966. Belgian engineers, who had to collect data from all ELDO members to design the telemetry system, were the first to confront the interface problem. They suggested that the Secretariat be given substantially more authority in a two-level management scheme. The first level would be a study bureau to establish specifications. It would be at the national level but under the "functional authority" of the ELDO Secretariat, and it would have authority to approve modifications and make technical decisions. Through control of the national bureaus, the Secretariat would impose consistent standards and processes. At a higher level in the organization, the Secretariat would have greater power, "corresponding in the English sense to the word 'control' (monitoring plus decision authority)." Belgian delegates proposed to staff this level with seventy engineers, with seven "inspector generals" from each national program under the direction of a management director. The engineers would focus on integration problems and look for future problems, while the seven inspector generals and the seven national program managers would meet with the Council to discuss problems at least every other month.[41]

Countering the Belgian proposal, the French proposed an Industrial Integrating Group that would exchange information among the government and industrial firms. The French solution provided information but did not give the Secretariat the power to enforce solutions. The Industrial Integrating Group would collect information and pass its recommendations to the Secretariat, which in turn could recommend changes with the member states and the ELDO Council. Perhaps not coincidentally, the Industrial Integrating Group would be led by SEREB, the French organization that coordinated the French rocket program.[42]

In matters political, French proposals carried more weight. German delegates supported the French because they did not want a strong project manager. Not wanting the project manager to have financial control, the Dutch supported the Germans. The ELDO Council decided to appoint a project manager for Europa I but to strictly limited the manager's authority. Council members directed, "The Project Manager shall remain within the approved technical objectives, timescale, programme cost to completion and total appropriations under each country chapter in the current budget." The manager would have to "pay due regard to the opinions and advice of other directors, but the decisions would be his own responsibility." The Council also required that he "act in agreement with the Member States concerned regarding budget transfers."[43] Without authority over budgets, the project manager could take no significant actions without agreement from the member states.

Along with appointing a project manager, the ELDO Council requested that the Secretariat investigate program management procedures and agreed with "the necessity of adopting a system for providing delegations and the Secretariat with continuous and full preventative information on the progress of the current programmes." Secretary-General Carrobio reported back, agreeing that a "Corps of Inspectors" should review ELDO and make recommendations concerning processes, structure, and management. Carrobio also proposed an "integrating group set up by industry and subordinate to the Secretariat's authority." Such a group would enhance ELDO's position.[44]

The French proposal prevailed. In July 1966, the ELDO Council approved the *Europa II* vehicle,[45] which could place a small communications satellite into geosynchronous orbit, and agreed to create an integration group, known as Société d'Étude et d'Intégration de Systèmes Spatiaux (SETIS [Company for the Study and Integration of Space Systems]), to strengthen the Secretariat. Beginning as a division of SEREB, SETIS had the same analysis and integration functions and was then spun off into a separate organization.[46]

SETIS had only advisory capacity, reporting to the ELDO Project Management Directorate, which the Council also created at that time. Under the new system, the Project Management Directorate assigned project managers to Europa I and Europa II. Each country selected its own project manager, who reported to the national organization and to the ELDO project manager, who distributed information to member states through the Scientific and Technical Committee. SETIS worked only on *Europa II* because *Europa I* was soon to begin integration testing. For Europa II, the Secretariat now had authority to place contracts directly with industry. ELDO's Europa I project manager remained virtually powerless.[47]

Secretary-General Carrobio expected that SETIS would strengthen ELDO's technical capabilities. Because the SETIS engineers came from Europa II contractors, SETIS would ensure better communication between ELDO and the manufacturers "by means of direct contacts." SETIS planned to hire forty engineers for the Europa II program and sixty more after that, arranged in three divisions: PAS Vehicle (*Europa II*) Development, Planning and Information, and System Integration. Despite SETIS's apparently broad charter, its power was strictly limited; it could not amend contracts or change costs, schedules, or technical performance except through the Secretariat. Because the Secretariat did not have much power in these matters, SETIS could only analyze information that member states and their contractors were willing to provide.[48]

In December 1966, the chairman of the Corps of Inspectors delivered his committee's report, which described "the problems of the interfaces" and "the role played in this matter by the Secretariat." The report noted, "The German Authorities and the German industrials repeatedly stressed the difficulties which have resulted from an incomplete solution of the interface problems, which they attribute to gaps in the methods of coordination." A number of interface problems bedeviled the project, and the Corps of Inspectors concluded that the Secretariat should define its methods and intentions to deal with interface issues.[49]

Carrobio responded the next month, first to specific problems noted by the Corps of Inspectors. The Secretariat had "played a determining role in reconciling the viewpoints" of the French and Germans in making design changes after a test failure in February 1966. However, in the case of problems with the German third stage, neither the German authorities nor ELDO could control foreign suppliers for third stage components. Even here, Carrobio stated, "These were not so much a matter of principles as of practical difficulties due to slippages in the development programme of Germany's foreign suppliers." Because German contractor Bölkow had no control over the suppliers, Carrobio argued, "It is then up to the Secretariat to intervene and endeavour to find and win acceptance for the least harmful compromise, and this has been done on all occasions." Carrobio believed that only minor fixes to ELDO's organization were necessary, not a complete overhaul.[50]

ELDO member state delegates took some time to approve the new processes and procedures proposed by the Corps of Inspectors. After an ELDO visit to NASA's Goddard Space Flight Center to learn more about project management techniques, the ELDO Council finally ratified the new program management procedures in September 1967. SETIS came into official existence on January 1, 1968. Both project management and SETIS strengthened ELDO's Europa II project. Europa I remained hampered by indirect contracting and a lack of authority.[51]

As ELDO, the national governments, and the contractors started to build and integrate *Europa I,* they found numerous communication and interface problems. Complaints bubbled up from the contractors through the national delegations to the ELDO Council, leading to an enhancement of ELDO's project management capability. The new procedures gave the Europa II project manager the authority to make direct contracts and gave him a staff that could monitor events more closely. However, even for Europa II, the Secretariat still had limited authority to modify contracts, costs, and schedules. Unfortunately, ELDO's immediate future hinged on Europa I and its less effective organization.

Disaster and Dissolution

To uncover technical problems prior to the upcoming test flights, ELDO contracted with Hawker Siddeley Dynamics to develop an electrical mockup of the *Europa I* and *Europa II* launchers at its facility in Stevenage. The company, in conjunction with engineers from other contractors, assembled complete stage mockups (sometimes without engines) prior to the F5 through F11 tests and flights. Hawker Siddeley ran several kinds of tests, including injection of faulty signals, electromagnetic interference, and flight sequence testing. These uncovered numerous design problems, which Hawker Siddeley then reported to the ELDO Secretariat and participating companies.[52] Although Hawker Siddeley engineers found numerous problems, they did not find enough of them.

ELDO's next major flight test, originally scheduled for 1965, mated France's *Coralie* second stage with Britain's *Blue Streak* first stage. After engineers aborted eight launch attempts because of various technical glitches and unfavorable weather, the ninth finally flew in August 1967, with the first stage operating properly. The second stage successfully separated, but its engines never fired, sending the stage crashing prematurely to Earth. Subsequent investigations showed that the problem stemmed from an electrical ground fault in the second stage, which deenergized a relay in the first stage, leading to a failure of the second stage sequencer, which then failed to issue commands for the second stage to fire. In short, the launch failed because of an electrical interface problem between the first and second stages.[53]

The next attempt with *Coralie* came in December 1967. It failed just as the first flight had, with its engines "failing to light up." In this case the failure occurred because of an electrical interface problem between the second stage and its connection to the ground system on the launch pad. Electromagnetic interference also hampered communication with the vehicle's safety system, causing the loss of flight data and the potential of an inadvertent explosion.[54]

After the loss of the first *Coralie* in August, the French considered second stage problems to be minor. However, with another failure of French equipment, French authorities reacted with urgency. In a major internal reorganization, the French contracted with SEREB to manage the second stage program. SEREB had been involved with the program until 1963, performing feasibility studies and initial designs. After that time, authority rested with the military Laboratory for Ballistic and Aerodynamic Research (LRBA) and the Bureau Permanent Nord Vernon. The French Space Agency now interposed SEREB between itself and LRBA to ensure closer surveillance and management of the program.[55]

SEREB proposed a major vehicle redesign to improve reliability. The proposals involved performing more qualification tests on *Coralie* components and replacing several major components with others that SEREB used in its *Diamant* rocket design. ELDO's technical group rejected the French proposals because they were very costly and, more importantly, because they would disrupt the entire program, including designs for the first and third stages.[56]

ELDO engineers supported LRBA rather than SEREB, rejecting SEREB's proposal to replace components. They stated, "The approach adopted by the French authorities is mainly due to the formation of a new technical direction team which is naturally anxious to use equipment with which it is familiar while being less familiar with the equipment it proposes to replace." ELDO engineers chided SEREB: It "will have to make great efforts to be as familiar with the programme as the present team—which has 'lived with' the *EUROPA I* launcher for five years and is at present very experienced—not only in order to develop its equipment but also to take account of the specific contingencies of ELDO." The member states rejected SEREB's proposal in favor of ELDO's proposal to upgrade and test existing second stage components.[57]

While the French regrouped, ELDO management assessed the impact of the project management reforms. The Secretariat divided its Project Management Directorate into two divisions, "one responsible for technical and time-scale aspects and the other for financial and contract aspects," while other directorates provided support to them. Secretary-General Carrobio noted, "The principle is now acknowledged, inside and outside the Secretariat, that additional work or modifications to approved work require the prior agreement of the project management directors." The coordination between member states and the Secretariat was improving, but still there were problems with schedule reporting (member states delivered PERT reports late) and cost control (member states did not thoroughly check contractor proposals). On Europa II, ELDO's system of monthly progress reports and meetings worked smoothly.[58]

In the next *Europa I* flight, in December 1968, ELDO engineers felt vindicated in their earlier resistance to SEREB's proposals because both *Blue Streak* and *Coralie* worked perfectly. Even though the German *Astris* exploded shortly after separation, ELDO and German engineers believed that they would isolate and repair the propulsion system problems they thought responsible.[59]

Their confidence was unfounded, for on the next attempt, in July 1969, *Astris* failed precisely as before, with an explosion within one second of separation from the second stage. The Germans realized, just as the French had two years before, that they had major problems. The Germans formed four committees: a government committee to investigate the failure, a committee to investigate the rest of the design, an internal committee of the contractor ASAT, and a committee to oversee and coordinate the other three committees. Contrary to expectations, the investigators found that the explosions resulted not from a third stage propulsion problem but from an electrical failure in the interface between the third stage and the Italian test satellite that ignited the safety self-destruct system. The Italians had already noticed sensitivity in these German circuits during their tests, but neither they nor the Germans recognized the importance of the finding.[60]

German engineers fixed the electrical troubles, but the third stage showed new problems in the last flight of *Europa I,* in June 1970. This time, the resulting investigation showed two third stage failures. First, an electrical connector disconnected prematurely, preventing separation of the Italian test satellite. Engineers traced this mechanical failure to the pressure between trapped air in the mechanism and the vacuum of space. Second, the third stage propulsion feed system failed, probably because of contaminants that kept a pressure valve open. These failures led the ELDO Council to create a quality assurance organization in 1970, but because of a lack of staff, it could not cover all sites and processes.[61]

In November 1971 *Europa II* flew for the first and last time. This vehicle, which included a new Perigee-Apogee stage, blew up two and one-half minutes into the flight because of an electrical malfunction caused by a failure of the third stage guidance computer. This too was an interface problem; the computer had been manufactured in Great Britain and delivered to German contractor ASAT, which took no responsibility for proper integration. At last, ELDO member states reacted strongly. As stated by General Aubinière, ELDO's new secretary-general, "The failure of F11 in November 1971 brought home to the member states—and this was indeed the only positive point it achieved—the necessity for a complete overhaul of the programme management methods."[62]

The ELDO Council appointed a committee to review every aspect of the program. The committee, which included senior engineers and executives from government and industry in the United States and Europe,[63] issued a devastating indictment of ELDO's organization and management, finding numerous technical problems that resulted from the lack of authority and inadequate communications. Poor electrical integration in the third stage was the immediate problem, having caused the failures of flights F7–F11. This was a function both of the poor management of the German company ASAT and of the contracting of the Secretariat, which did not assign integration authority to any of the contractors working on third stage components. Failures resulted from interface problems with components delivered by foreign contractors to ASAT, and between ASAT's two partners, MBB and ERNO.

Communications between ASAT and its two partners were extremely poor. ASAT was a small organization created by the German government solely to coordinate the large companies MBB and ERNO on the ELDO third stage. Communications between ASAT and other firms supplying third stage components were even worse, leading to a design that obeyed "none of the most elementary rules concerning separation of high and low level signals, separation of signals and electrical power supply, screening, earthing, bonding, etc." This made the British guidance computer and sequencer extraordinarily sensitive to noise and minor voltage variations, which in turn caused it to fail in the F11 flight.[64]

By far the most significant recommendations were to abolish indirect contracting and to ensure the definition of clear responsibilities for interfaces. Member states at last gave the Secretariat the authority to place contracts. Now desperate for solutions, ELDO adopted a number of American techniques, including the full adoption of phased planning, work breakdown contractual structures, and preliminary and final design reviews for the Europa III program, ELDO's hoped-for improvement to Europa II.[65]

General Aubinière, the new secretary-general and former director of the French Space Agency, hoped that stronger project management would turn ELDO around. Unfortunately, ELDO never got another chance. After the British withdrew financial support in 1971, the remaining partners had to develop a new first stage or purchase *Blue Streak* stages. Disillusioned after the F11 failure, the Germans threatened to withdraw. With continuing political disagreements over launch vehicles and over cooperation with the United States space shuttle program, ELDO's support evaporated. The member states eliminated Europa II while the F12 launch vehicle was on its way to Kourou in April 1973. ELDO bid for a part in the American shuttle program, but when the Americans withdrew its proposed Space Tug, a vehicle to boost payloads to higher orbits, ELDO's time was up. The member states dissolved it in February 1974.[66] Twelve years of negotiation, compromises, and struggle came to an end.

Conclusion

Political and industrial interests drove the formation of ELDO, Europe's largest cooperative space project. All of the major powers preserved national interests through indirect contracting, by which national governments maintained authority. Even after being strengthened in 1966 by adding a Project Management Directorate and an integrating organization, SETIS, the Secretariat was limited to collecting information and distributing it through the Technical Directorate to the ELDO Council. Member states compounded the Secretariat's weakness by changing objectives, as British support weakened and French leaders lobbied to create a more powerful rocket that could boost communications satellites to geostationary orbit.

ELDO's weakness resulted in a series of failures caused by interface problems. The ELDO Secretariat could neither create nor enforce consistent documentation, processes, or quality. Nor could it force contractors to communicate with each other. These problems resulted in badly designed electrical circuitry between the British, French, and German stages as well as internal to the German third stage because of the poor management and attitude of German and British contractors. Six consecutive failures were the result, all but one because of interface failures traceable to poor communications between member countries and contractors.

While rocket and space programs in the United States and the Soviet Union all confronted numerous failures in the process of learning how to build these complex technologies, inadequate organizational structures and processes compounded ELDO's problems. Like James Burke, the Jet Propulsion Laboratory's first Ranger project manager, Europa I project managers had no control over major elements of the project. ELDO's unbroken series of failures mirrored Ranger's early problems, showing the criticality of organizational issues. Europeans had the capability to build rockets, as shown by successes with Nazi Germany's *A-4* in World War II, and the postwar British *Blue Streak* and French *Veronique* and *Diamant.* However, all of these projects benefited from strong, centralized organizations and close working relationships between government and contractors to ensure better communication. The ill-fated *Europa* launchers had none of these advantages.

ELDO combined many of the worst management ideas into a single, pitiful organization. Its engineers, managers, and directors struggled against a fatally flawed management structure that was almost the exact antithesis of systems management in the United States. Where systems management promoted strong authority for the project manager, in ELDO the manager's authority was virtually nil. Systems management required critical attention to interfaces, but ELDO initially ignored them; no single individual or group ever analyzed *both* sides of the interface to ensure compatibility. Component quality assurance—through inspections, testing, and documentation—was standard in the United States but only randomly present in ELDO. ELDO's hapless record and defective structure was a warning to European leaders that cooperative technology development required true cooperation. Europeans would begin launcher development again, but this time on a much sounder basis. The new effort would build upon Europe's successful science satellite group, ESRO.

SEVEN

ESRO's American Bridge across the Management Gap

> Firms have shown themselves anxious to collaborate with ESRO as a means of gaining useful experience of the newer management techniques which are indispensable for the effective control of the financial as well as the technical aspects of large and complex projects.
>
> —J. J. Beattie and J. de la Cruz, 1967

The European Space Research Organisation (ESRO) presented a welcome contrast to the ongoing embarrassments of the European Space Vehicle Launcher Development Organisation (ELDO). Created as a service organization for European space scientists, ESRO overcame its initial organizational difficulties and developed a successful series of scientific satellites. Its achievements proved that effective European space cooperation was possible. Although ELDO had been the Europeans' prime organization to develop space technology, its failure paved the way for ESRO to become the route of choice across the management gap between the United States and Europe.

ESRO's success owed a great deal to its greater contractual authority (compared to ELDO) and to American assistance. While European industry and powerful interest groups focused on the military and economic significance of launchers, ESRO's scientific satellites seemed insignificant. ESRO's Convention and procedures consequently had fewer provisions to protect national economic interests than did ELDO, giving ESRO authority that ELDO never had. In addition, whereas Americans did not want to aid Europeans in rocketry or communications satellites, they cheerfully gave technical and financial assistance to European science.

These factors help explain ESRO's rise from a small service organization to the core of Europe's integrated space organization. Through the authority of its Convention, the ability of its engineering and managerial staff, and the help of the United States, ESRO and its descendant, the European Space Agency (ESA), mastered the art of systems management.[1]

The Inception of ESRO

The creation of ESRO began with the activities of Edoardo Amaldi, Italian physicist and one of the founders of the Conseil Européen pour Recherche Nucléaire (CERN [European Committee for Nuclear Research]). In the summer of 1958, after a conversation with his friend Luigi Crocco, a rocket propulsion expert and professor in Princeton University's Department of Aeronautical Engineering, Amaldi proposed a European space program modeled on CERN. The new space organization should have high goals, Amaldi said, comparable to efforts in the United States and the Soviet Union, but have "no connection with whatsoever military agency." He believed that it should be "open, like CERN, to all forms of co-operation both inside and outside the member countries."[2]

Amaldi learned from Crocco and from American aeronautical engineer Theodore von Kármán some difficulties in modeling a European space organization on CERN. Because the military had developed virtually all rockets, excluding the military would be difficult. Crocco also believed that it would be difficult to convince European parliaments to spend the huge sums necessary for space-based science research. Von Kármán thought it necessary to include the military at the beginning to jump-start the civilian effort. He suggested working through the North Atlantic Treaty Organization. Amaldi demurred and eventually found a strong ally for his purely scientific space organization in his friend Pierre Auger, a French physicist and CERN ally.[3]

When Amaldi contacted Auger in February 1959, Auger was organizing the French Committee for Space Research. Auger was supportive of Amaldi's proposal and suggested the French organization as a model. French scientists and administrators were considering a two-phase program: a small initial effort based on sounding rockets, and a more ambitious program to include satellite launches and lunar or solar probes. After the two men met in April 1959,

Amaldi helped establish an Italian space research committee on the French model. Amaldi also sent a paper titled "Space Research in Europe" to prominent scientists and science administrators in Western Europe.[4]

These contacts led to an informal meeting of scientists from eight different countries at Auger's Paris home in February 1960. At the next meeting, held in April 1960 at the Royal Society in London at the behest of the British National Committee for Space Research, the British presented their extensive space research plans and the possibility that the British government might offer the *Blue Streak* rocket as the basis for a European launcher. Auger hosted the next meeting in Paris in June 1960 to consider "A Draft Agreement Creating a Preparatory Commission for European Collaboration in the Field of Space Research."[5] During the second Paris meeting, British delegates removed launchers from discussion because of negotiations under way between the British and French governments concerning the use of the *Blue Streak*. With launcher considerations eliminated, the scientists and scientific administrators focused on creating a European space research program using sounding rockets and satellites.[6]

Further discussions clarified the purpose and scope of ESRO and established goals for its initial scientific program and facilities. ESRO would support space scientists throughout Europe. It excluded launch vehicles, although at the request of the Belgian delegation, it did include the development of supporting technologies. ESRO planners envisaged a two-phase effort: an initial program using sounding rockets, and a more advanced program of sophisticated scientific satellites.

Bruising negotiations determined the sites of ESRO facilities. To expedite coordination with ELDO, ESRO's headquarters wound up in Paris. ESRO's most important facility was its engineering unit to develop spacecraft and integrate scientific experiments, the European Space Technology Centre (ESTEC). Originally located in Delft, The Netherlands, ESTEC soon moved to the small coastal town of Noordwijk, north of The Hague. The telemetry data analysis center went to Darmstadt, West Germany, the sounding rocket range to Kiruna, Sweden, and a small science research center to Delft. A new scientific research center with ill-defined functions, located near Rome, satisfied Italian demands for an ESRO facility. In 1967 ESRO officials moved satellite tracking to Darmstadt, where combined with the data analysis center it became the European Space Operations Centre. ESRO established remote tracking stations in Alaska, Norway, Belgium, and the Falkland Islands.[7]

European scientists originally conceived of ESRO as an organization run by scientists, for scientists, on the model of CERN. CERN provided an infrastructure for European physicists to perform experiments with particle accelerators. In CERN's organization, scientists determined the technical content of projects and infrastructure, and ran daily affairs. Administrators had little control over CERN's funding, and significant overruns developed.

ESRO provided a similar service function to space scientists through provision of sounding rockets, satellites, and data collection and analysis facilities. Scientists selected ESRO's experiments, but, unlike in CERN, engineers developed and operated the infrastructure. The British insisted on strong financial controls, ensuring that if ESRO overran its budget, it would cut projects instead of forcing governments into funding overruns.[8] Because the founding scientists did not want ESRO's scientific expertise to rival that of the member states, they restricted ESRO's scientific research capabilities, making its engineering character more pronounced. ESRO's engineering culture made it a very different organization from CERN.

Ten countries signed the ESRO Convention of June 1962: the United Kingdom, France, Italy, West Germany, Belgium, The Netherlands, Sweden, Denmark, Spain, and Switzerland. ESRO came into official existence on March 20, 1964, with Pierre Auger as secretary-general.

Organizing ESRO's Early Projects—with American Help

ESRO selected projects in consultation with scientific groups, a council representing the national governments, and its own scientists. Ad hoc groups recommended experiments to the Launching Programmes Advisory Committee (LPAC), which in turn selected a few of them to form a satellite payload. The LPAC recommended payloads to the Scientific and Technical Committee and the Administrative and Financial Committee. These committees then presented their assessments to the ESRO Council, which made the final decision. The Council passed its decision to ESRO headquarters, which then authorized ESTEC and the other ESRO organizations to begin work.[9]

Unlike ELDO's, ESRO's authority included contract placement and control. The ESRO Convention required that ESRO "place orders for equipment and industrial contracts among the Member States as equitably as possible, taking into account scientific, technological, economic and geographical considerations." To do so, ESRO created a register of member state suppliers. For items costing more than 10,000 French francs, ESRO's financial rules required that ESRO request bids from industry, unless ESRO had "no alternative but to go directly with one supplier." ESRO submitted all purchases of greater than 500,000 French francs to its Administrative and Finance Committee, along with any purchases outside of the member states.[10]

Although ESRO's day-to-day affairs revolved around engineering, scientists heavily influenced the selection of projects and experiments. The short-term sounding rocket program consisted of seventy-one launches from Sardinia, Norway, Sweden, and Greece between 1964 and 1968. For the medium term, ESRO's satellite program consisted of two spin-stabilized scientific spacecraft, known as *ESRO-I* and *ESRO-II.* Shortly thereafter, ESRO approved three more satellites: a polar orbiting satellite known as the *Highly Eccentric Orbit Satellite* (*HEOS-A*) and two complex attitude-stabilized spacecraft known as *Thor-Delta 1* and *Thor-Delta 2* (*TD-1, TD-2*).[11]

In 1963, scientists and administrators in ESRO's Preparatory Commission initiated internal and contract feasibility studies for ESRO-I. Performed early in 1964, these contract studies contributed to the definition of the scientific payload. ESRO released its tender for ESRO-I in November 1964. After ESTEC engineers evaluated the resulting proposals, ESRO awarded several contracts in April 1965. The Laboratoire Central de Télécommunications of Paris received the contract for project management and satellite integration, and companies in Switzerland and Belgium received "associate" contracts.[12] Each of these companies had subcontractors, including some American companies offering components not readily available in Europe, such as sun sensors, batteries, and test equipment.[13]

ESRO-II evolved at the same time—and with the same process. ESTEC scientists and engineers began internal design studies in July 1963 and awarded external design study contracts to a Belgian firm and a Swiss university.[14] ESTEC engineers deliberately introduced variations in the designs that these institutions studied so as to assess different methods of attitude control.

After completion of these feasibility studies, ESTEC engineers wrote technical specifications used in the call for tenders in June 1964. In November, ESRO selected British firm Hawker Siddeley Dynamics as prime contractor, and Hawker Siddeley subcontracted to several British and French companies.[15] ESTEC let separate contracts for the command, telemetry, and checkout subsystems and also coordinated the "supply of sub-systems to the prime contractor." Hawker Siddeley had responsibility for project management, specifications, interfaces, structures, and integration.[16]

The HEOS project started somewhat later and evolved similarly. In early 1964, a study group rejected a planetary mission because it would have required the construction of large ground stations. Instead, the group recommended a spacecraft in a highly eccentric orbit around Earth. ESRO endorsed the project in July 1964, at which time ESTEC appointed a project manager. ESTEC conducted feasibility studies in late 1964 and issued calls for tender in June 1965. In November, ESTEC awarded the contract to a consortium led by Junkers Corporation.[17] The Junkers team hired Lockheed Space Corporation from the United States to provide consulting and to supply high-reliability parts. Development began in January 1966. The HEOS project marked the first contract award to a consortium, a trend that would soon become the norm for European industry. Following American trends, it also marked the first use of an incentive contract instead of a cost-plus-fixed-fee contract.[18]

ESRO-I and ESRO-II took advantage of the National Aeronautics and Space Administration's (NASA's) offer to launch ESRO's first two satellites free of charge. *HEOS-A* also used an American launcher, but ESRO had to pay for the service. NASA offered its junior partner technical assistance, including project training, reviewing test results, participating in joint reviews, conducting launch operations, and supplying additional tracking and data acquisition support. Goddard Space Flight Center (GSFC) managed NASA's contributions. Through working groups and design reviews, GSFC space scientists and engineers guided ESRO personnel through their early projects.[19]

What did ESRO administrators, scientists, and engineers learn from GSFC personnel? GSFC managers began projects by issuing a project specification and a competitive tender. They expected the prime contractor to issue a spacecraft handbook for experimenters and to attend monthly interface meetings with experimenters and other organizations. Cost-plus-fixed-fee contracts were the norm for development; administrators monitored them through monthly contractor reports. GSFC managers stressed the importance of change control, coordinating all design changes with contributing organizations. The initiator of changes had to submit a written proposal to the project manager, who had final authority.[20]

GSFC and ESRO formed joint working groups for ESRO-I and ESRO-II so that ESRO personnel could learn from their NASA counterparts, so that NASA personnel could learn about European methods, and so that solutions for common problems and interfaces could be worked out. NASA provided representatives from its technical divisions, along with the project manager and representatives from *Scout* launch vehicle contractor Ling-Temco-Vought. The working groups covered topics such as mechanical and electrical interfaces, launch and mission procedures, reliability and quality assurance, and testing and verification. The Europeans heeded American advice regarding interfaces, iteratively defining and reworking interfaces until they were consistent across subsystems and between the spacecraft and the launch vehicle.[21]

High-level ESRO administrators visited the United States in 1964. ESTEC's technical director, chairman of the Scientific and Technical Committee, and Large Satellites Division chief visited NASA headquarters, GSFC, Princeton University, and Grumman Corporation to learn about the organizational and technical aspects of the Orbiting Astronomical Observatory project. In February 1965, the ESRO-I project manager and scientists visited Rice University in Houston. After visiting Rice—and presumably NASA officials from the Manned Spacecraft Center—they visited renowned space scientist James van Allen of Iowa State University.[22]

With little spacecraft experience, European contractors also used American assistance when they could get it. ESRO-II prime contractor Hawker Siddeley had "a considerable amount of technical liaison" with Thompson-Ramo-Wooldridge (TRW). Junkers hired Lockheed as a technical and management consultant for HEOS and to procure high-reliability components. These supplemented other European-American industrial interactions at that time, which included Boeing's one-third purchase of Bölkow, TRW's establishment of Matrel Corporation with Engins Matra, North American's cooperation with Société d'Études de la Propulsion par Réaction, and Douglas Company's cooperation with Sud Aviation.[23]

British organizations also assisted ESRO. A visit by ESRO administrators to the U.K. Ministry of Aviation focused on financial estimating and reporting procedures and the use of the Program Evaluation and Review Technique (PERT). On its projects, the Ministry of Aviation placed contracts for the entire development and planned future expenditures by acquiring predicted financial profiles from its contractors. The ESRO visitors found that the ministry and some of its contractors used PERT/TIME for schedule planning. Because PERT was available in Britain only through International Business Machines (IBM) computers and produced summaries intended for "PERT oriented managements which are even rarer in the U.K. than PERT oriented project teams," the ministry recommended that PERT was not a good solution for scheduling and cost-estimating problems.[24]

ESRO's inexperienced project personnel depended on contractors. According to ESRO-II project manager Ants Kutzer, one important innovation was to have ESRO representatives attend all project meetings between its two major contractors, Hawker Siddeley and Engins Matra. He stated that "although unusual . . . the most valuable aspect . . . was that the ESTEC project team gained detailed technical knowledge of the design as well as experience."[25]

Kutzer was an acute student of research and development (R&D) management, having read American studies of R&D contracting, including those by RAND and the Harvard Business School that documented American missile management methods. He followed the development of scheduling tools such as PERT as well as early systems engineering texts. To Kutzer, the lesson of these early studies and tools was that for complex projects, managers needed to deploy new methods that identified "all of the activities required to meet the end objective." These methods should, Kutzer said, show complex interrelationships and constraints, including interfaces; predict the time and cost outcome; optimize resource allocation; and be flexible enough to adapt to rapid change.[26]

Because of the great diversity in nationalities involved in the ESRO-II project, Kutzer believed that it needed new management techniques. He emphasized close coordination and communications between ESRO, GSFC, and the contractors. He felt that "informal exchange of ideas and techniques" in the NASA-GSFC working group and numerous subgroups made "a major contribution to project success." Kutzer discussed formal specifications and documents at regular meetings and supplemented them with informal meetings. To minimize the effect of "rather exhaustive listening to a foreign language," Kutzer systematized meeting agendas to standardize the vocabularies used in the meeting. So too did ESRO-I managers.[27]

The HEOS program borrowed extensively from American management models, resulting in thorough advanced planning, stronger project management and systems engineering, and the development of European consortia. Junkers led the winning industry team, drawing extensively on Lockheed for management ideas. Lockheed helped to bring together the Europeans' diverse companies and traditions in the process of developing the proposal bid:

> The firms had mutually coordinated their bid proposals in Europe and afterwards met in Sunnyvale to write the definitive bid text. In these weeks, very lively discussions with the experienced specialists of the American firm led to strong contact between the executives of the European firms, which became decisive for cooperation in the realization of the project. Furthermore, the participants learned to link the same ideas with the same words . . .
>
> The consortium's bid consisted of approximately 1,000 pages, around one-third of which concerned management and cost-estimating questions. Without the advice of the American firm, this part in particular would not have undergone such a deep treatment.[28]

The Junkers team bid far surpassed earlier and contemporary bids in the detail and attention given to management. Junkers won the HEOS contract by a considerable margin. Junkers team members believed that they won by such a wide margin primarily because "it could be assumed [by ESRO] with great certainty that the bidders had constructed quite realistic time and cost plans."[29] For later ESRO projects, European teams adopted the Junkers approach, including using American consultants, constructing detailed management plans, and employing close-knit consortia to carry those plans out.

Both ESRO and its contractors experimented with PERT and other planning techniques to determine their utility for spacecraft projects. Europeans learned of PERT through American papers and contacts and acquired it through the use of IBM computers. As an experiment, the ESRO-I prime contractor, the Laboratoire Central de Télécommunications (LCT), proposed

HEOS spacecraft. On the HEOS project, European contractors formed their first consortium based on recommendations received from U.S. contractor Lockheed. Courtesy NASA.

using IBM PERT/Cost software. LCT management found the reports generated very useful for analyzing completed and future activities and expenses. They delivered reports every three months to ESRO, including a cost plan, a bar chart for management, and a detailed cost report. Because ESTEC did not have PERT but wanted its own PERT plan for top-level project events, ESTEC managers updated their own network by hand from LCT's PERT results. The ESTEC project team also generated weekly bar charts. Near the end of the program, as the spacecraft progressed in a serial fashion through testing, the project stopped using PERT, switching to simple bar charts.[30]

HEOS prime contractor Junkers developed sophisticated PERT networks, using a detailed monthly cycle for acquiring inputs and generating outputs. In part because Junkers's incentive contract rewarded a launch in early 1969, Junkers emphasized the use of PERT to control schedules. It created an 800-event network for HEOS, backed by a system of Planning Change Notices that tied PERT to engineering and management changes. As LCT had done on ESRO-I, Junkers produced bar charts for managers and more detailed network listings for planners, and it also used PERT/Cost with generally favorable results.[31]

ESRO-II management also used PERT through prime contractor Hawker Siddeley but paid more attention to developing new techniques to measure project progress and to implement configuration control. Project manager Kutzer recognized that although configuration control as used by the U.S. Air Force was useful for ESRO-II systems engineering, he could not implement it, because of the lack of experience and lack of detailed requirements for ESRO-II. Instead, ESTEC engineers established the requirements through the "unusual approach" of attending all technical and contractor meetings. They limited themselves to being "technically suspicious and taking nothing for granted," and they tried to be "pessimistic about success and to find weak links," to ensure strong testing, and "to support the contractors."[32]

Hawker Siddeley's project manager developed a new process to assess progress during specification development. He created an empirical method whereby planners gave each proposed specification a "marks loading," a numerical value that depended upon the amount of work expected. The engineer responsible for the specification could estimate the percentage of work completed against the specification. For example, a specification worth 50 marks loading and estimated at 60% complete would be given a current marks value of 30. By adding the total of all current marks values and dividing this sum by the total marks loading for the project, Hawker Siddeley acquired an estimate of the amount of work completed and the amount remaining.[33]

After completing the specifications and establishing a design baseline, Kutzer and Hawker Siddeley's managers implemented a configuration control process. They developed standardized forms that summarized subsystem status, including acceptance test status, reliability, defect reports, modifications, and information and action items still required. When the subsystem successfully passed its tests and supplied the relevant paperwork, ESTEC issued a Design Acceptance Note that formally accepted the subsystem. After issuance of the Design Acceptance Note, engineers could modify the design only by submitting a Modification Proposal Authorization Form. It included the modification and the reasons for it; the estimated cost, schedule, and weight impact; and its effect on other subsystems, documentation, and firms.[34]

One European deficiency was the lack of environmental test facilities suitable for satellite checkout. Europeans knew from American published papers and personal contacts that satellites had to be thoroughly tested on the ground, including vibration testing, charged particle radiation testing, and thermal vacuum testing. ESRO's initial program included substantial investments in facilities, including environmental test facilities. By 1966, ESTEC managers had two vacuum chambers and vibration systems under construction. In 1966, ESTEC used its own vibration system and vacuum facilities to test the ESRO-I structural test and thermal models. Prior to completion of ESTEC's facilities, ESRO and its contractors used British, French, and American facilities.[35]

Largely because of their lack of environmental test facilities, European companies did not have parts that met the high standards typically associated with American satellites. All three of ESRO's initial satellites procured high-reliability electronic components from the United States.[36] When American companies could not deliver these scarce components on schedule, delays of several months ensued for ESRO-I and HEOS. Only the ESRO-II program avoided significant delays in procurement of high-reliability American parts.[37]

Each project acquired American expertise through direct consultation and interaction with GSFC personnel. During a design review by GSFC personnel in October 1966, NASA experts stated that ESRO had not sufficiently accounted for the space thermal environment and needed to perform further analysis and testing. In response, ESRO created a complex thermal model and added a test in a French thermal vacuum chamber, both of which verified the adequacy of the original design. NASA reviewed *ESRO-I* launch operations plans in October 1966. After the Flight Readiness Review at ESTEC from August 12–16, 1968, ESRO managers and engineers waxed enthusiastic: "It was a great moment for the Project Team, when at the end of the Flight Readiness Review, the NASA experts declared *ESRO-I* flight ready."[38] GSFC experts performed similar reviews for *ESRO-II* and *HEOS* between 1966 and 1968.[39]

After some initial problems, ESRO's satellites operated successfully. *ESRO-II* launched in May 1967 but never made it into orbit, as NASA's *Scout* launcher exploded during ascent. ESRO regrouped and successfully launched a second model in May 1968. *ESRO-I* successfully launched in October 1968, and *HEOS* in December.[40]

ESRO personnel began their first projects recognizing their own inexperience and took advantage of NASA's offer to help, both in launching their first two satellites for free and in training ESRO personnel in spacecraft design and management. European managers, engineers, and scientists visited the United States to learn American methods, and their American counterparts reciprocated by visiting ESTEC during working group meetings, design reviews, and Flight Readiness Reviews for ESRO's satellite projects. GSFC personnel gave substantial help to ESRO, as did American contractors TRW and Lockheed to ESRO's prime contractors Hawker Siddeley and Junkers for ESRO-II and HEOS. ESRO and its contractors used American models for its testing programs, planning methods, configuration control, and reliability assessment. They also acquired and used PERT with the help of IBM computers. On HEOS, Lockheed advised European contractors to emphasize management issues, leading to a strong consortium that won the bid by a large margin. The Junkers consortium's successful bid was the model for contractor consortia on later bid opportunities. The technical success of the satellites ESRO launched in 1968 and 1969 showed the value of ESRO's methods.

ESRO's Crisis

Despite significant technical progress, in 1967 ESRO was an organization in crisis, for financial, scientific, technical, and political reasons. The member states kept a short rein on ESRO's finances, even tightening their grip as time passed. Scientists fought among themselves about which ESRO satellite programs would be considered top priority, and program costs escalated. ESRO's administrative structure made decisions difficult, and its satellite operations were awkwardly organized. Finally, the emergence of communications satellites was a catalyst to expand ESRO's mission from pure science to commercial technology development. Between 1966 and 1968, ESRO and the member states confronted these issues, leading to significant changes in ESRO's role and organization.

Although always troubled, ESRO's financial status worsened in 1966. Member states controlled ESRO's budget by setting three-year and eight-year caps. During the first three-year period, ESRO underspent its financial cap by 120 million French francs because it did not build facilities as rapidly as planned. Much to the surprise of ESRO's administrators, scientists, and engineers, the ESRO Council refused to carry the funds forward to the next three-year period, 1967–69. Because ESRO planners had assumed that they would be able to carry these funds forward, ESRO's programs were in jeopardy. Shocked administrators canceled ESRO's most expensive program, the Large Astronomical Satellite. In addition, ESRO reduced the three planned TD missions to two, then eventually one.[41]

These reductions exacerbated scientific struggles over payloads and experiments. ESRO could not satisfy the atmospheric researchers, astronomers, geophysicists, and cosmic ray physicists competing to fly experiments. Sounding rockets were cheap enough to fly frequently, but spacecraft were a different story. In the end, scientists flew fewer experiments than they desired and had to take turns with complex missions.[42]

Overruns on ESRO's early projects contributed to ESRO's cost problems. ESRO-II project manager Kutzer estimated his project's cost overrun at 50% with a schedule slip of 10%. In addition to this, ESRO had to build a second *ESRO-II* satellite because of the loss of the first in a launcher failure. HEOS performed better from a cost standpoint. Its project manager estimated the increase of the prime contract at 18% after accounting for inflation, with a 13% schedule delay. He considered this excellent and credited it to ESRO's authority to choose contractors based on factors other than the cheapest bid. The Junkers team did not have the cheapest bid, but it did have by far the most detailed one. Cost increases, along with the loss of carryover funds, resulted in an immediate cut in current projects and pressure to cut future ones.[43] The first new project to feel ESRO's pressure to accurately predict and control costs was the Thor-Delta 1 and Thor-Delta 2 (TD-1/2) project.

Based on the successful example of the Junkers team for HEOS, most firms organized themselves into consortia for the February 1967 TD-1/2 contract award. In the design competition, contractor cost estimates ranged from 99 million to 176 million French francs. With such widely varying estimates, ESRO managers and engineers could not predict the final cost. ESRO eventually selected the MESH consortium, consisting of Matra, Entwicklungsring Nord (ERNO), Saab, and Hawker Siddeley.[44] ESRO management soon rued this selection, as MESH's cost estimates grew dramatically, even during negotiations. Technical problems of three-axis stabilization led to these ballooning costs, which induced ESRO management to cancel the preliminary contract in mid-1968 and reduce the two-spacecraft program to a single satellite, *TD-1*.[45] By that fall, ESRO reentered into a contract with MESH for a single vehicle with a simplified stabilization system.[46]

The economic potential of communication satellites also confronted ESRO. In the early 1960s, the United States' Echo, Telstar, and Syncom projects demonstrated the reality of satellite communications. The United States led efforts to create Intelsat, a semiprivate organization to develop commercial satellite communications. To prepare for Intelsat negotiations, the Europeans organized the Conférence Européene de Télécommunications par Satellites (CETS [European Conference for Satellite Telecommunications]), which, in turn, contracted with ESRO to investigate communications satellite design. At the same time, the French and Germans developed their own bilateral program, known as Symphonie, and Italy started its own program, known as Sirio. By 1968, the CETS effort through ESRO focused on television broadcasting in conjunction with the European Broadcasting Union. ESRO's new director-general, the British scientist Hermann Bondi, concluded that politically "ESRO could not survive on a very narrow base of pure scientific research." He resolved to convince his scientific colleagues of that fact.[47]

In the fall of 1967, ESRO management proposed to manage the CETS program as it had its earlier projects, by giving out a number of associate contracts, one with the integration task. This was no longer acceptable to European industry. Having whetted their appetites on ESRO's scientific satellites, and sensing the possibility of commercial gain on a larger scale, the contractors put pressure on ESRO to let a single prime contract. As stated by CETS spokesmen, "Although the advantages of the ESRO proposition have been recognized—in particular the flexibility in choosing contractors, the control of program costs, and the geographic distribution of contracts—certain delegations expressed very clearly the opinion that industry should be conferred global system responsibility, because this task permits them to acquire a highly profitable experience in the domain of technical management and finance of complex projects." ESRO management caved in and gave industry the prime contractor role. However, ESRO maintained the tasks of preparing specifications, defining the entire system (including ground system and infrastructure), and providing detailed supervision of performance, cost, and schedule. The prime contractor prepared system specifications and approved subsystem designs in collaboration with ESRO but otherwise managed, integrated, and tested the satellite.[48]

Big projects such as the Large Astronomical Satellite held another danger for ESRO. On this program, the British and French national delegations insisted on contracting through their national organizations, as they did in ELDO. Only project cancellation saved ESRO from this dangerous precedent, which might have doomed ESRO to ELDO's fate.[49]

Uneven contract distribution also created political problems for ESRO. As French leaders had planned, France won a significant percentage of technical and facility contracts, with ESRO headquarters in Paris, and contracts for ESRO's first three satellites. The Netherlands, with ESTEC on its soil, also did well. British and Italian leaders complained bitterly to the ESRO Council, demanding that contracts be distributed to more closely match contributions to the organization.

The loss of carryover funds from ESRO's first three years of operation caused a major financial crisis for ESRO, leading to several project cancellations. Although its cost overruns were smaller than those of many comparable American spacecraft projects, ESRO's stringent financial rules amplified their effect. Technical troubles on TD-1 led to further cost increases, adding more pressure. When combined with the growing importance of nonscientific applications like satellite telecommunications, pressure grew among the member states and within ESRO itself for changes in its goals and management. In response, the ESRO Council commissioned internal and external reviews to improve ESRO's organization and management.

Reforms and Results

Dutch physicist J. H. Bannier, former chairman of the CERN Council, headed ESRO's external review. Bannier's commission recommended strengthening project management by giving project managers control over technical specialists and control over expenditures, "within well defined limits, in the same way as . . . the technical aspects of the work" were controlled. The commission also recommended that ESRO have more responsibility to issue and monitor contracts to relieve the Administrative and Financial Committee of trivial duties. It also advised separating policy decisions at ESRO headquarters from day-to-day project management and technical duties at ESTEC. Director-General Bondi defined new financial rules in November 1967 to ensure a minimum 70% return of funding to member countries from ESRO contracts. By 1968, ESRO had implemented the bulk of the commission's recommendations.[50]

Financial problems with the TD project and industry's assertiveness spurred ESRO's internal review, which focused on improvements to project implementation. Director-General Bondi initiated "urgent actions," leading to the development of a working paper by Mr. Schalin, an ESRO headquarters administrator. Schalin's paper, presented to the ESRO Council in March 1968, presented a blueprint for improving procedures for "forecasting, preparing for and implementing major projects in ESRO." Basing his paper on "previous efforts and a recent visit to NASA," Schalin proposed changes to project cost estimation, contracting procedures, and project control.[51]

To improve cost estimation, Schalin proposed a version of NASA's phased project planning, along with project cost-estimating formulas that GSFC administrators were developing. Schalin noted that on TD-1/2, ESRO and the MESH consortium did not develop reliable cost estimates until one year after contract award. On other contracts, a stable estimate did not occur until 75% of the funds had been committed. Schalin proposed that ESRO spend 5–10% of the total project cost for project definition and design phases to develop accurate cost estimates prior to final contract decision. He recommended adoption of Project Definition and Detailed Design Definition phases between feasibility studies and the development contract award. Both phases would use competitive study contracts lasting six to twelve months, using either fixed-price or cost-plus-fixed-fee contracts.

Schalin realized that past experience was the most reliable guide for early forecasts. With little experience, ESRO's forecasts were problematic. NASA, on the other hand, had developed spacecraft cost-estimating formulas with a purported accuracy of 10–20%. ESRO administrators evaluated cost-estimating techniques from GSFC and the Illinois Institute of Technology Research Institute, finding that ESRO spent a higher proportion of funding on administration and less on hardware than did GSFC. This surprised ESRO administrators, who had expected that their costs would be less than those of the United States. Instead, this "Atlantic factor" represented ESRO's more difficult communication problems.[52]

Finally, Schalin recommended substantial changes to ESRO's project control methods. He noted that HEOS-A had developed extensive project control procedures, which TD-1/2 implemented only after a huge underestimate. Schalin proposed a stronger contract support organization, in conjunction with rigorous change control based on detailed specifications, implemented through a network plan with work breakdown structures tied to the accounting system. Schalin's paper became the starting point for ESRO's project management reforms.[53]

By fall 1968, the Bannier committee's organizational reforms and Schalin's recommendations took effect. ESRO managers restructured the budget and introduced a new financial plan. ESTEC managers found that for current projects, "*HEOS A2,* being essentially a repeat, and the Special Project [TD-1], being further advanced, cannot be fully adapted to the new procedures." For these, ESTEC increased the project manager's authority with "extra staff controlling cost and contract matters." It also provided more technical support to project teams and implemented a "combined network/work package/cost control system."[54]

After considering the "possibility for the Organisation taking over the prime contractorship from industry," TD-1 managers added twenty-three full-time personnel to the project, including eight for a new Project Control Section, bringing the total to fifty. In addition, TD-1 used part-time staff from the technical divisions—the equivalent of forty full-time employees.[55] Managers divided the MESH consortium's tasks into 600 work packages, monitored on a monthly cycle of "data collection-processing-presentation-interpretation-decision." For HEOS-A2 and ESRO-IV, which were follow-up projects to HEOS and ESRO-II, ESTEC managers used phased planning, although without competition.[56]

Phased planning crystallized into full-fledged procedures in June 1969, with the release of ESRO's *Phased Planning for Scientific Satellite Projects Guidelines.* ESRO management intended phased planning to provide "clear and logical build-up stepwise of information for management decisions" that maximized project insight, yet minimized project commitment up to the decision points. The guidelines defined six phases: mission studies, preliminary feasibility studies, project definition and selection, design and detailed definition, development, and operations. For each phase, ESRO defined specific processes and products, which organization was to produce them, and who was to decide whether to proceed. By November 1969, ESRO's Administrative and Finance Committee translated the phased planning guidelines into new procedures for satellite contracts. ESRO used the full phased planning procedure on its next satellite, *COS-B.*[57]

A second major initiative sponsored by Director-General Bondi was the development of a management information system (MIS). In August 1968, Bondi approved a proposal to establish an MIS Study Group. The study group's efforts converged with ongoing activities at ESTEC to develop a project control system.[58]

Two MIS models stood out as particularly relevant: the Centre National d'Études Spatiales (CNES, the French space agency) chart room, and the NASA Marshall Space Flight Center (MSFC) MIS. The CNES chart room was entirely manual, where the chart room itself was the "data bank" of historical, statistical, operational, and project information. MSFC's system was an

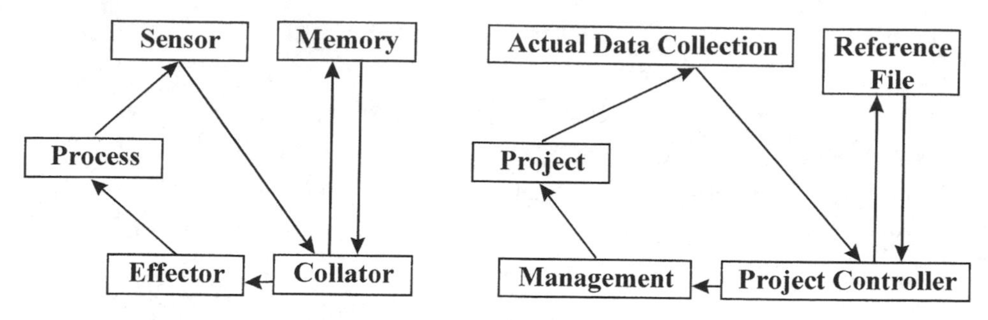

Hoernke's analogy of engineering and project control.

example of a fully computerized MIS that handled inventory, PERT project information, parts and reliability data, and videotape and library collections.[59]

To help determine what kind of system would be appropriate, MIS Study Group member H. Hoernke compared the concept of project control to the more general concept of control in engineering. He defined five components of an engineering control system: the process to be controlled; the sensor; the collator, which compares "what *is* taking place with what *should* be taking place"; the memory, which records the standard for what should be happening; and the effector, which changes the process toward the standard. On projects, various organizations performed the functions of the sensor, collator, memory, and effector. In his analogy, the project control group acted as the collator of data collected by various teams, and management acted as the effector.[60]

By November 1968, ESTEC personnel decided to use the IBM PMS 360 program for project control.[61] This restricted management information to items compatible with the PMS program, including cost and schedule information stored in work breakdown structures. The IBM program could print a number of standard reports, which Hoernke described in his assessment. In that same month, Hellmuth Gehriger of the ESTEC Contracts Division proposed to extend his division's work to include "management services." In early 1969, ESTEC management approved his proposal to create a Management Services Section, which would perform research on managerial problems, recommend standards, keep statistics on management performance, and aid ongoing programs. Gehriger also proposed that this group, eventually called the Project Control Section, perform operations research and systems analysis studies.[62]

The MIS Study Group found numerous cases of redundant information generation, haphazard use, inconsistent levels of detail, and widely varying implementation of procedures and automation within ESRO. It concluded that creating an MIS would be a formidable task but provide substantial benefits. Aside from standard arguments that an MIS would improve management performance and organizational efficiency,[63] the MIS group also argued for an MIS for political reasons. Foreseeing a possible merger with ELDO and CETS, the group deduced, "When two or more organisations merge, it is clearly the one that has the more organised structure that is at an advantage." The group stated, "To control its contractors, ESRO must at least be as efficient as industry in managing the information problem." It thought "ESRO should play a leading role in getting industry used to advanced management techniques." In the opinion of group members, an MIS would give ESRO a distinct advantage in the coming bureaucratic battle.[64]

With Bondi's endorsement, the MIS Study Group decided to develop a "semi-integrated system" where each ESRO facility would have its own computer system. This had the advantage of virtually ensuring the "rationalisation of information" at that site but the disadvantage of potentially promoting information barriers between ESRO sites. A fully centralized system at Paris, the group believed, would be very complex and would politically generate "high resistance," both from ESRO personnel at other facilities and from the national delegations.[65]

The resulting distributed computer system, called the Planning, Management, and Control (PMC) System, began operation in January 1971. At the beginning of a project, managers and engineers entered financial and schedule information, coded as work package numbers tied to the accounting system. The system generated reports, including internal budgets, plans, differences between plans and actual events, and changes to plans.[66]

At ESTEC, the Project Control Section of the Contracts Division ran the PMC System. The section's director, Hellmuth Gehriger, became a vocal managerial theorist. He believed that project control was rapidly developing into a science. The project manager determined what had to be "project controlled," and the project controllers determined how to manage the projects. Gehriger used the critical path method to schedule tasks in the phased planning cycle and produced planning and control documents and information

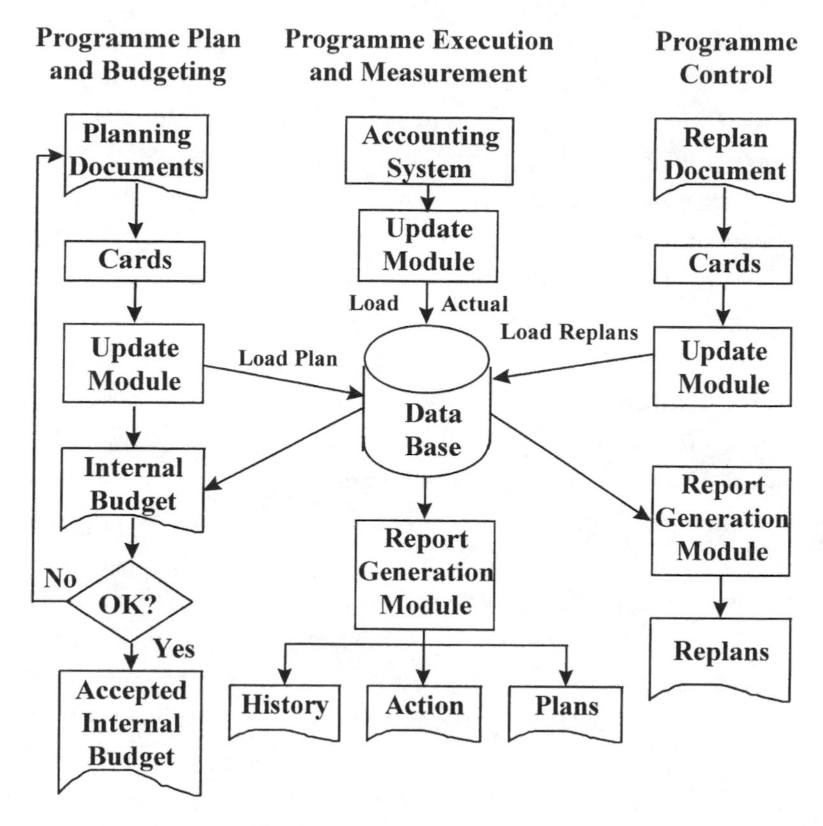

The ESRO Planning, Management, and Control System.

flows to closely monitor project cost and schedule. From reported status information, project control personnel prepared a Key Event Schedule Trend Analysis report that monitored schedule trends for slippage, a sign of impending technical and cost difficulties. Under Gehriger's guidance, the Project Control Section developed a sophisticated management scheme that adapted American managerial concepts to the European context.[67]

ESRO management did not give its project managers complete control. In December 1969, the head of ESTEC's Satellites and Sounding Rockets Department proposed that ESTEC assign all project personnel to his department, thereby giving him and the project manager control over all project personnel. ESRO Director of Administration Roy Gibson refused. Although Gibson agreed that ESTEC's use of manpower was "extravagant," he did not "recommend the same cure." Gibson believed that the proposal would "lead to the

Satellites and Sounding Rockets Department becoming an independent state within a state." Instead, Gibson argued for better coordination of manpower within a matrix system. This accorded with the opinion of the head of the Personnel Department, who noted that the "home division" contained the "reservoir of knowledge" required for projects.[68]

Over the next four years, ESRO managers struggled with the division of authority between the project manager and the technical divisions. ESTEC established a new Programme Coordination Division to coordinate personnel between the projects, the technical divisions, and the European Space Operations Centre. After difficult negotiations, ESTEC Director Hammarström issued a directive that required project managers to create and review a support plan twice each year and to request support from technical divisions through standardized forms.[69]

ESRO's financial and political crises of 1967 and 1968 spurred a series of organizational reforms. Under pressure from national delegations and Director-General Bondi, ESRO managers followed a path well trodden by the U.S. Department of Defense and NASA. ESRO adopted phased planning to provide management decision points and better cost estimates. For the TD project, it was too late to implement phased planning, so instead ESRO created the Project Control Section and implemented configuration management techniques, which ESRO used on all later programs. By 1971, the ESRO MIS was partially operational at ESTEC under Gehriger's Project Control Section. ESRO personnel looked forward to the impending merger with ELDO and CETS, believing they had the organizational advantage.

The merger would soon take place, spurred by ELDO's failure as well as the opportunities and hazards of the American shuttle program. On Spacelab, the Europeans' contribution to the shuttle, the new ESA changed from being NASA's junior apprentice to being NASA's partner. To make this partnership work, NASA would require even further "Americanization" of European management methods.

ESA, Spacelab, and the Second Wave of American Imports

By 1972, European space programs were in the most severe of their many political crises. ELDO's first *Europa II* launch had failed, and an international commission was investigating the organization's many flaws. German and Italian leaders backed away from their commitments to ELDO. The United States offered the Europeans a part in the new shuttle program as a way to cut development costs and to discourage the European launcher program. In addition, the United States sent ambiguous signals to European governments on its willingness to launch European communications satellites. All European governments agreed on the criticality of satellite telecommunications, but ESRO's charter did not allow commercial applications, and the telecommunications organization CETS had no satellite design capability. A sense of crisis pervaded negotiations among the European nations concerning their space programs.

Each of the member states of ELDO, ESRO, and CETS confronted the myriad problems and opportunities differently. The United Kingdom decisively turned its back on launch vehicles in favor of communications satellites. British leaders believed it wiser to use less expensive American launchers, so as to concentrate scarce resources on profitable communications satellites. Doubting that the United States would ever launch European satellites that would compete with the American satellite industry, French leaders insisted on developing a European launch vehicle. German leaders became disillusioned with launchers and were committed to cooperation with the United States. Because of their embarrassing failures on ELDO's third stage, they wanted to acquire American managerial skills. Italy wanted to ensure a "just return" on its investments in the European programs by having a larger percentage of contracts issued to Italian firms. For smaller countries to participate in space programs, they needed a European program.

The result of these interests was the package deal of 1972 that created the ESA. Britain received its maritime communications satellite. France got its favorite program, a new launch vehicle called *Ariane.* Germany acquired its cooperative program with the United States, *Spacelab,* a scientific laboratory that fit in the shuttle orbiter payload bay. The Italians received a guarantee of higher returns on Italy's investments. All member states had to fund the basic operations costs of the new ESA and participate in a mandatory science program. Beyond that, participation was voluntary, based upon an à-la-carte system where the countries could contribute as they saw fit on projects of their choice. European leaders liquidated ELDO and made ESRO the basis for

ESA, which came into official existence in 1975. ESTEC Director Roy Gibson became ESRO's new director-general and would become the first leader of ESA.[70]

Two factors drove the further Americanization of European space efforts: the hard lessons of ELDO's failure, and cooperation with NASA on Spacelab. By 1973, both ELDO and ESRO had adopted numerous American techniques, leading to a convergence of their management methods. On the Europa III program, ELDO finally used direct contracting through work package management, augmented by phased planning to estimate costs. ESRO managers had done the same in the TD-1 program and in ESRO's next major project, COS-B. ESRO's two largest new projects, Ariane and Spacelab, inherited these lessons.

For Ariane, ESRO made CNES the prime contractor. CNES had responsibilities and authority that ELDO never had. ESRO required that CNES provide a master plan with a work breakdown structure, monthly reports with PERT charts, expenditures, contract status, and technical reports. CNES would deliver a report each month to ESRO and would submit to an ESRO review each quarter. CNES could transfer program funds if necessary within the project but had to report these transfers to ESRO. ESRO kept quality assurance functions in-house to independently monitor the program. Reviewers from ESRO required that CNES further develop its initial Work Breakdown Structure, which was "neither complete nor definitive," and specify more clearly its interface responsibilities. The French took no chances after the disaster of ELDO and fully applied these techniques, as well as extending test programs developed for ELDO and for French national programs such as Diamant. French managers and engineers led the Ariane European consortium to spectacular success by the early 1980s, capturing a large share of the commercial space launch market.[71]

While Ariane used some American management methods,[72] Spacelab brought another wave of American organizational imports to Europe. This suited Spacelab's German sponsors, who wanted to learn more about American management. NASA required that the Europeans mimic its structure and methods. ESRO reorganized its headquarters structure to match that of NASA, with a Programme Directorate at ESRO headquarters, and a project office at ESTEC reporting to the center director. The NASA-ESRO agreement required that ESRO provide a single contact person for the program and that this person be responsible for schedules, budgets, and technical efforts. Contrary to its earlier practice, ESRO placed its Spacelab head of project coordination with the technical divisions such as power systems, and guidance and control. ESRO used phased project planning, eventually selecting German company ERNO as prime contractor. ERNO contracted with American aerospace company McDonnell Douglas as systems engineering consultant, while NASA gave ESRO the results of earlier American design studies and described its experience with Skylab.[73]

To ensure that the program could be monitored, NASA imposed a full slate of reviews and working groups on ESRO and its Spacelab contractors. NASA held the Preliminary Requirements Review in November 1974, the Subsystems Requirements Review in June 1975, the Preliminary Design Review in 1976, the Critical Design Review in 1978, and the Final Acceptance Reviews in 1981 and 1982. Just as in the earlier ESRO-I and ESRO-II projects, NASA and ESRO instituted a Joint Spacelab Working Group that met every other month to work out interfaces and other technical issues. Spacelab managers and engineers organized working sessions with their NASA counterparts, including the Spacelab Operations Working Group, the Software Coordination Group, and the Avionics Ad Hoc Group. European scientists coordinated with the Americans through groups such as the NASA/ESRO Joint Planning Group. The two organizations instituted joint annual reviews by the NASA administrator and the ESRO director-general and established liaison offices at each other's facilities.[74]

ESRO did not easily adapt to NASA's philosophy. From the American viewpoint, ESRO managers could not adequately observe contractors' technical progress prior to hardware delivery, so NASA pressed ESRO to "penetrate the contractor." NASA brought more than 100 people from the United States for some Spacelab reviews, whereas in the early years of Spacelab, ESRO's entire team was 120 people. Heinz Stoewer, Spacelab's first project manager, pressed negotiations all the way to the ESRO director-general to reduce NASA's contingent. According to Stoewer, NASA's presence was so overwhelming that "we could only survive those early years by having a close partnership between ourselves and Industry."[75]

NASA engineers and managers strove to break this unstated partnership and force ESA[76] managers to adopt NASA's approach. During 1975, NASA and ESA engineers worked on the interface specifications between *Spacelab* and the shuttle orbiter. During the first Spacelab preliminary design in June 1976, NASA engineers and managers criticized their ESA counterparts for their lack of penetrating questions. To emphasize the point, NASA rejected ERNO's preliminary design. ESA finally placed representatives at ERNO to monitor and direct the contractor more forcefully.[77]

One of the major problems uncovered during the Preliminary Design Reviews was a large underestimate of effort and costs for software development. These problems bubbled up to the ESA Council, leading to a directive to improve software development and cost estimation. ESTEC engineers investigated various software development processes and eventually decided to use the NASA Jet Propulsion Laboratory's software standard as the basis for ESA software specifications. ESA arranged for twenty-one American software engineers and managers from TRW to assist. TRW programmers initially led the effort but gave responsibility to their European counterparts as they acquired the necessary skills. With this help, software became the first subsystem qualified for Spacelab.[78]

Another critical problem was Spacelab's large backlog of changes, which resulted from a variety of causes. Organizational problems magnified the technical complexity of the project, which the Europeans had underestimated. One major problem was that ESA allowed ERNO to execute "make-work" changes without ESA approval. ESA, whether or not it agreed with a change, had to repay the contractor. Another problem was the relationship between ESA and NASA. The agreement between the two organizations specified that each would pay for changes to its hardware, regardless of where the change originated. For example, if NASA decided to make a change to the shuttle that required a change to *Spacelab,* ESA had to pay for the *Spacelab* modification. This situation occurred frequently at the program's beginning, leading to protests by ESA executives.[79]

Spacelab managers solved their change control problems through several methods. First, they altered their lenient contractor change policy; ESA approval would be required for all modifications. Second, ESA management resisted NASA's continuing changes. In negotiations by Director-General Roy Gibson, NASA executives finally realized the negative impact that changes to the shuttle had on *Spacelab*. As ESA's costs mounted, NASA managers came to understand that they could not make casual changes that affected the European module. The sheer volume of changes to *Spacelab* reached the point where ERNO and ESA could not process individual changes quickly enough to meet the schedule. In a "commando-type operation," ESA and ERNO negotiated an Omnibus Engineering Change Proposal that covered many of the changes in a single document. ERNO managers decided to begin engineering work on changes before they were approved, taking the risk that they might not recoup the costs through negotiations.[80]

Technical and managerial problems led also to personnel changes at ESA and ERNO. At ESA, Robert Pfeiffer replaced Heinz Stoewer as project manager in March 1977, while Michel Bignier took over as the program director in November 1976. At ERNO, Ants Kutzer replaced Hans Hoffman. Kutzer, who was Hoffman's deputy at the time, was well known as the former ESRO-II project manager and a promoter of American management methods. After his stint at ESRO, Kutzer became manager of the German Azur and Helios spacecraft projects. For a time, ESA Director-General Roy Gibson acted as the de facto program manager because of Spacelab's interactions with NASA, which required negotiations and communications at the highest level.[81]

ESA and ERNO eventually passed the second Preliminary Design Review in November 1976 by developing some 400 volumes of material. The program still had a number of hurdles to overcome, of which growing costs were the most difficult. Upon ESA's creation, the British insisted that ESA, like ESRO, have strong cost controls. For à-la-carte programs such as Spacelab, when costs reached 120% of the originally agreement, contributing countries could withdraw from the project. Spacelab was the first program to pass the 120% threshold. After serious negotiations and severe cost cutting, the member states agreed to continue with a new cap of 140%. Eventually, ESA delivered Spacelab to NASA and the module flew successfully on a number of shuttle missions in the 1980s.[82]

The transfer of American systems engineering and project management methods to ESA was effectively completed by 1979, with the creation of the Systems Engineering Division at ESTEC, headed by former Spacelab project manager Stoewer. According to Stoewer, one of the System Engineering Division's main purposes was to ensure "more comprehensive technical support to mission feasibility and definition studies." The Systems Engineering Division established an institutional home for systems analysis and systems engineering, which along with the Project Control Division ensured that ESA would use systems management for years to come.[83]

Conclusion

ESRO began as an organization dedicated to the ideals of science and European integration. The CERN model of an organization controlled by scientists, for scientists, greatly influenced early discussions about a new organization for space research. However, space was unlike high-energy physics in the diversity of its constituency, the greater involvement of the military and industry, and the predominance of engineers in building satellites. Engineering, not science, became ESRO's dominant force.

From the start, ESRO personnel looked to the United States for management models. Leaders of ESRO's early projects asked for and received American advice through joint NASA-ESRO working groups for ESRO-I and ESRO-II. The United States offered ESRO two free launches on its *Scout* rockets but would not allow the Europeans to launch until they met American concerns at working group meetings, project reviews, and a formal Flight Readiness Review. On HEOS, executives and engineers from the loose European industrial consortium led by Junkers traveled to California for advice, where Lockheed advised much closer cooperation, along with a detailed management section in its proposal. ESRO's selection of the Junkers team by a large margin made other European companies take notice, leading to tightly knit industrial consortia that at least paid lip service to strong project management.

When ESRO's member states refused to let ESRO carry forward its surplus in 1967, they immediately caused a financial crisis. Significant cost overruns on TD-1 and TD-2, along with strong pressure to develop telecommunications satellites by issuing a prime contract to industry, spurred ESRO to beef up its management methods. ESRO adopted from NASA models phased planning, stronger configuration control, work package management, and an MIS at each ESRO facility.

American influence persisted through the 1970s. On Spacelab, NASA imposed its full slate of methods onto ESA. Some, such as contractor penetration, ESA initially resisted, before technical problems weakened the Europeans' bargaining position. ESA managers and engineers eagerly adopted other methods, such as software engineering, once they realized the need for them. These became the standards for ESA projects thereafter.

Europeans created an international organization that developed a series of successful scientific and commercial satellites, captured the lion's share of the world commercial space launch market, and pulled even with NASA in technical capability. To do so, they adapted NASA's organizational methods to their own environment. In the process, ESRO changed from an organization to support scientists, into ESA, an engineering organization to support national governments and industries. ESA's organizational foundation, like that of NASA and the air force, was systems management.

EIGHT

Coordination and Control of High-Tech Research and Development

When we mean to build, we first survey the plot, then draw the model; and when we see the figure of the house, then we must rate the cost of the erection: which if we find outweights ability, what do we then but draw anew the model in fewer offices, or at last desist to build at all?

—William Shakespeare, *King Henry IV*, Part 2, Act 1, Scene 3

Systems management was a typical product of the Cold War, consisting of organizational structures and processes reflecting the interests and expertise of the social groups that created it. Facing intense pressure to deliver state-of-the-art technologies on tight schedules, military officers, managers, scientists, and engineers contributed their respective types of expertise and vied for control of the development process. Competition and cooperation both flourished in the pressure cooker of the early Cold War, but ultimately these groups formed a coherent process for the development of large-scale technologies.

A common thread was the emphasis on systems. To Bernard Schriever and other air force officers, the "systems approach" meant unifying the research and development (R&D) command structure, to unite in one organization what the air force had traditionally accomplished in separate organizations. To engineers at the Jet Propulsion Laboratory (JPL), the systems approach meant accounting for operations and logistics in a missile's design. To RAND analysts, the systems approach meant applying mathematical techniques to a larger set of technologies and organizations than previously considered. In each case, "systems" implied an expansion of capabilities, authority, and concepts beyond what was traditional.

Each social group developed means of communication and control to enhance its effectiveness and authority. Managers and military officers developed communication procedures that funneled information to a central point and disseminated decisions and authority from that point. Less obviously, working groups of scientists and engineers also channeled information and authority, but in their case to their own working groups. Based upon the needs of the Cold War, each group used systems rhetoric to gain authority, then designed "procedural systems" to keep it.

Scientists and engineers first developed systems analysis and systems engineering to analyze and coordinate the development of large-scale technologies. Organizing through ad hoc committees typical of academia, these technically competent individuals maintained power through the informality of their communication, which seemed unique to each problem and project. Standardization did not seem possible, and technical experts wanted to keep it that way.

Those in control of the project funding and goals thought otherwise. Military officers and managers sought ways to control the seemingly uncontrollable R&D process and ultimately found a solution in configuration management, which linked managerial hierarchies with technical committees. Through the configuration control board (CCB), managers used the "power of the purse," requiring scientists and engineers to give cost and schedule estimates with each design change. This gave them a proxy measurement to assess technical progress and hence assess the scientists and engineers as well.

Changes were inevitable in complex ballistic missile and spacecraft designs. Their novelty meant that technical and managerial teams learned as they went. Much of this learning came through failure, when missiles exploded and spacecraft failed far from Earth. Because many, if not most, problems encountered were interface problems traceable to communication problems, or manufacturing defects traceable to simple errors in repetitive processes, organizational means were primary in eradicating these errors. Systems management significantly improved missile and spacecraft reliability.

The Cold War provided the context and motivation for military and civilian authorities to fund scientists and engineers to develop complex, heterogeneous weapons systems. In short, scientists and engineers working on Cold

War military projects created technical coordination processes that managers and military officers appropriated and modified to control R&D. Just as scientific management enabled managers and engineers to coordinate and control factory workers in the first decades of the twentieth century, systems management enabled military officers and civilian managers to coordinate and control scientists and engineers.

Social Groups, Values, and Authority

Alliances between scientists and military officers had grown during World War II on the Manhattan Project, in the Radiation Laboratory, and in operations research groups. The Cold War furthered this military-scientific partnership. Appealing to the imminent Soviet threat, military officers like Bernard Schriever promoted the systems approach in weapons development to quickly design and manufacture novel weapons such as ballistic missiles. Working with his scientific allies, Schriever built an organization initially run by military officers and scientists. Similarly, Army Ordnance officers allied themselves with JPL's research engineers to develop the *Corporal* missile. Both Army Ordnance and Schriever's "Inglewood complex" spent immense sums of money in concurrent development, designing, testing, and manufacturing missiles as rapidly as possible. The result for *Atlas* and for *Corporal* was the same: a radically new, expensive, and unreliable weapon.

Prematurely exploding missiles created a spectacle not easily ignored. JPL's engineering managers resolved to improve on their ad hoc methods and employed the systems approach their next missile, the *Sergeant.* The air force's next-generation missile was the *Minuteman,* on which Col. Samuel Phillips developed the system of configuration control to better manage costs and schedules. Both second-generation weapons were far more reliable than their predecessors, partly because of the switch to solid-propelled engines, and partly because of changes in organizational practices.

Over time, social measures of success changed. Initially, simply getting a rocket off the ground was a major accomplishment. Eventually, however, Congress expected that its large appropriations would buy technologies that worked reliably. Soon thereafter, congressional leaders wanted accurate cost predictions so they could weigh alternative uses for that money. Cost overruns came to be seen as failures. This was particularly true in Europe, where leaders promoted space programs mainly to spur economic development.

The Ranger program and its aftermath illustrated the power of Congress to change organizations. Under William Pickering's guidance, JPL used a loose matrix structure where most authority resided with the technical divisions. When Ranger's failures exposed JPL's organizational flaws, Congress required JPL to strengthen project management and implement more stringent procedures. Pickering and JPL's engineers resisted these changes, but Ranger's failures weakened their credibility. When National Aeronautics and Space Administration (NASA) Administrator James Webb threatened to remove all future programs from JPL, Pickering had little choice. He gave in. Similar pressures influenced the air force in the early 1960s and the European Space Research Organisation (ESRO) in the late 1960s. Systems management was the end result in each case. The European Space Vehicle Launcher Development Organisation's (ELDO's) attempts to strengthen project management did not succeed, because of weaknesses inherent in its authorizing Convention and the uncooperative attitude of its member states.

The first figure illustrates the relationships between the four social groups. In the early Cold War, military officer-entrepreneurs and scientists provided the authority and methods. I distinguish here between those military officers such as Bernard Schriever who promoted new systems, and others, such as Samuel Phillips, who brought them to fruition. Schriever acted in an entrepreneurial fashion and Phillips as a classical manager. In the air force, this period lasted from roughly 1953 until 1959, the heyday of the *Atlas* missile, before its many test failures led to change. Schriever acted as a visionary entrepreneur, albeit in an unconventional blue uniform. JPL's period of military entrepreneur-scientific control came from 1944 to 1952, when JPL's research engineers developed *Private* and converted *Corporal* from research to production. In both cases, expensive, unreliable weapons led to a concentration on cost and dependability for the next missiles, leading to approaches based on engineering and managerial values. Jack James at JPL and Phillips of *Minuteman* typified the no-nonsense managers that demanded dependability. Unlike engineers focused on research, such as Caltech's von Kármán and Malina, most engineers focused on the design and development of technological sys-

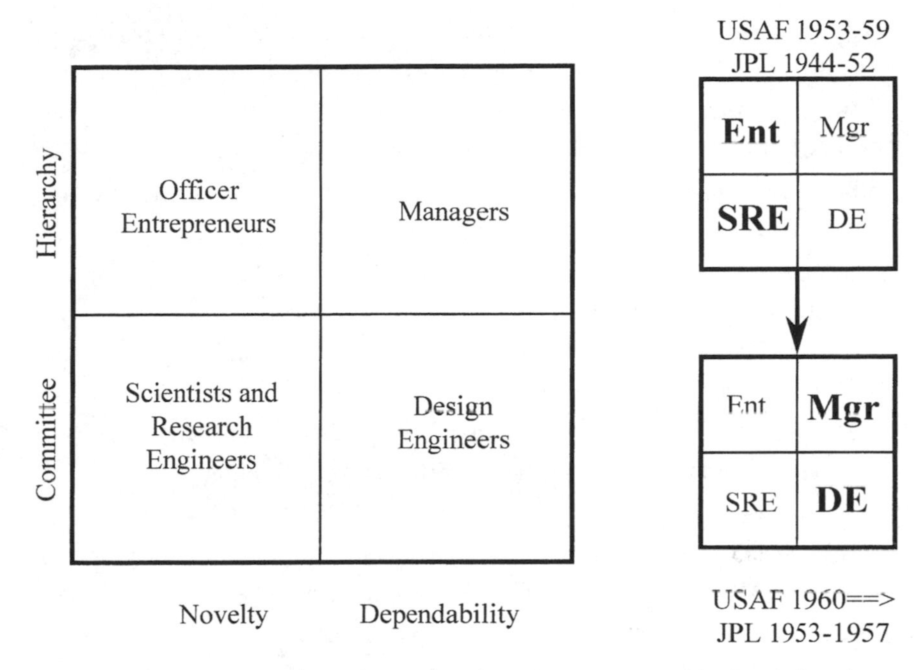

Cold War social groups and alliances. At JPL from 1944 to 1952, and in the air force between 1953 and 1959, entrepreneurial military officers and scientists (along with research engineers) formed a social alliance to promote novel weapons. After these periods, managers in the military and industry formed an alliance with design engineers to control costs and build dependable systems.

tems. For them, creating a product that worked was more important than creating one that was new.

NASA's history differed somewhat from that of the air force, because in the early years of NASA, the scientists and engineers controlled their own projects and methods. At JPL, the research-based methods prevailed in the laboratory's early years, and again later when Pickering shifted the laboratory into the space program, and satellite launches (and failures) were frequent as JPL raced with the clock and the Soviets.[1] JPL's new projects reverted to ad hoc committees to get fast results. Similarly, former National Advisory Committee for Aeronautics researcher Robert Gilruth of the Space Task Group ran the early manned programs with little interference from NASA Administrator Keith Glennan. Engineers and scientists made decisions locally in a highly decentralized organization. After the Ranger fiasco at JPL, and after the arrival of George Mueller and Samuel Phillips in the manned programs, NASA's high-level managers and engineers quickly centralized authority. A similar story was unfolding at ESRO, originally conceived of as an engineering service organization for scientists. By 1967, commercial interests predominated and European governments changed ESRO into an engineering organization run by managers to ensure cost control.

It is more difficult to determine who, if anyone, ran ELDO. With an ambassador as secretary-general and economic nationalism the primary driving force, ELDO did not have a single dominant group, one could argue. Neither engineers nor scientists controlled the organization. Nor could managers foster the communication necessary to break national and industrial barriers. If ever there existed a purely political organization for technology development, ELDO was it.

Each social group promoted its characteristic methods. Military officers and industrial managers promoted project management to centralize authority and used contractor penetration to closely monitor industry. Both groups also used competition to keep contractors honest and developed cost prediction and control methods such as configuration management and work package management. Scientists promoted analytic techniques such as statistical analysis of reliability, network mathematics, and game theory. Engineers used

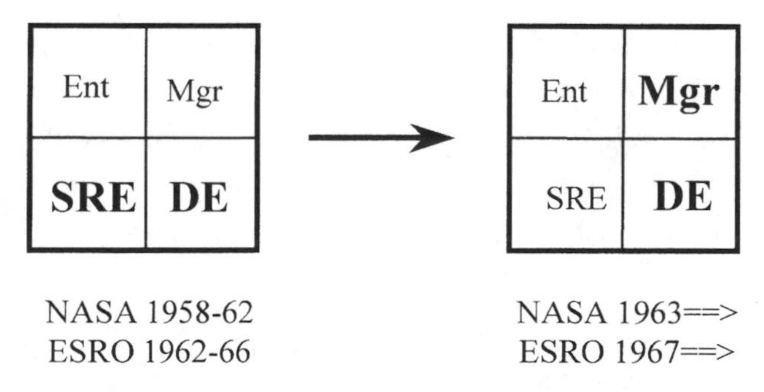

Authority changes at NASA and ESRO. In early NASA and ESRO, scientists and research engineers were allied with design engineers to build new technologies. After several years, both organizations shifted to more predictable development, with managers and engineers controlling events.

Scientists	Engineers	Military Officers	Managers
Operations Research	Systems Engineering	Concurrency	Phased Planning
Systems Analysis	Systems Integration	Technical Direction	PERT / Critical Path Method
Cost-Benefit Analysis	Contractor Penetration	Control Rooms	Contractor Penetration
Functional Organization with Committees	Functional Organization with Committees	Project Organization	Project and Matrix Organizations
	Design Freeze Change Control	Configuration Management	Configuration Management
Quantitative Reliability	Qualitative Reliability	Multiple Source Competition	Work Package Management
	Quality Assurance		Quality Assurance
	Environmental and System Testing	System Testing	

Systems management methods classified by the social groups that promoted them.

a variety of testing methods, inspection and statistical methods for quality assurance, and design freeze to stabilize designs.

Ultimately, systems management is a stable system because its methods and processes maintain roles for each of its constituent social groups. For systems management to remain stable over many years and projects, it had to have mechanisms for its constituent social groups to effectively interact. In the end, the primary mechanism became configuration management.

Committees, Hierarchies, and Configuration Management

Between 1962 and 1965, NASA's organization changed from a series of engineering committees to a mixture of committees overlaid with a managerial hierarchy. Similarly, between 1950 and 1964, JPL's committee structure gave way to hierarchical project management. Schriever's entrepreneurial Western Development Division also used ad hoc committees from 1953 to 1955, separated from the rest of the air force hierarchy. From 1956 onward, the air force hierarchy increasingly asserted control. These shifts signified changes in the balance of power between the hierarchical models of organization used by military and industrial managers, and the informal committees used by engineers and scientists.

The novel technologies of the 1950s required the services of scientists and engineers, who through their monopoly on technical capabilities influenced events. Schriever's alliance with scientists in the 1940s and 1950s brought scientists to the forefront of the air force's development efforts. In NASA's first few years, engineers at the field centers effectively controlled NASA and its programs. In both cases, scientists and engineers extensively used committee structures to organize activities. These working groups generated and used the information necessary to create the new technologies. Knowing that Congress would pay the bills, scientists and engineers essentially ignored costs. Indeed, if they could have made correct cost predictions, these would to some extent have invalidated their claim to be creating radically new technologies. Schriever shared the scientists' "visionary" bias. He argued for radical new weapons and the methods to rapidly create them by reminding everyone of the Soviet threat.

By the 1960s, the need for radical weaponry declined. When in 1961 reconnaissance satellites showed the missile gap to be illusory, arch-manager Robert McNamara, the new secretary of defense, immediately imposed hierarchical structures and centralized information systems to assert control. Similarly, the air force asserted control over Schriever's organization when reliability problems led to embarrassing questions from Congress. Ironically, the methods used by the air force and the Department of Defense to control Schriever were the methods that Schriever's group had created to control ballistic missiles. NASA's turn came after 1963, as Congress clamped down in response to NASA's wildly inaccurate cost estimates. Air force R&D veterans Mueller and Phillips imposed hierarchy and information systems over NASA's engineering committees.

NASA's early history showed that committees could successfully develop reliable technologies, but only when given a blank check and top priority. On the manned space flight programs, NASA's engineers and contract personnel had ample motivation. With clear goals and a national mandate, formal control mechanisms were unnecessary. Informal methods worked well both inside and outside NASA, as NASA engineers exerted firm control of contractors through informal but extensive contractor penetration. As long as Congress was willing to foot the bill for huge overruns, NASA's committees sufficed. When motivation was overwhelmingly positive, goals were clear, and funding was generous, coordination worked.

The history of ELDO illustrates how critical motivation and authority are to an organization's success. ELDO's primary function was to coordinate several national programs through committees whose only authority was their ability to persuade others. Unfortunately for ELDO, the national governments and industrial companies were at least as concerned with protecting their technologies from their national and industrial partners as they were with cooperation. By 1966, both the national organizations and ELDO began to recognize problems with this situation, and they created an Industrial Integrating Group to disseminate information. Without authority, neither the integrating group nor ELDO could bridge the communication gaps between contractors, leading to a series of interface failures and ultimately to ELDO's demise. Without motivation, authority, or unitary purpose, ELDO failed.

The trick to designing new technologies within a predictable budget was to unite the creative skills of scientific and engineering working groups with the cost-consciousness of managerial hierarchies. JPL and the air force developed the first link: configuration control. In both organizations, configuration control developed as a contractual association between the government and industry. Industrial contractors already used the design freeze as the breakpoint between design and manufacturing. Configuration control in the air force linked the design frozen by the engineers to the missile as actually built. When managers found that a number of missile failures resulted from mismatches between the engineering design and the "as launched" missile, air force managers implemented a system of paperwork to link design drawings to specific hardware components.

At JPL, configuration control developed because JPL designed the *Sergeant* missile, while industrial contractor Sperry was to build it. Deputy Program Director Jack James realized Sperry needed design information as soon as possible, so he required JPL engineers to document and deliver information in several stages. At each stage, James and others integrated the various engineers' information into a single package. James then controlled design changes by requiring engineers to communicate with him before making changes. This gave James the opportunity to rule on the necessity of the change and to ensure communication with other designers to coordinate any other implications of the change. This "progressive design freeze" worked so well that James imported it into his next project, Mariner. James and other managers expanded the concept on Mariner and its successors to include cost and schedule change estimates with every technical modification, thus tying costs and schedules to technical designs.

The air force also realized that tying cost and schedule estimates to engineering changes was a way to control engineers. Air force managers and engineers from The Aerospace Corporation expanded the concept to include the development of specifications. Soon the air force made specifications contractually binding and tied specification changes to cost and schedule estimates, just as it did for design drawings and hardware. This system of change control for a hardware configuration, tied to cost and schedule estimates, became known as configuration management. Minuteman program director Phillips recognized configuration management as an important tool, and he imposed it on Apollo to coordinate and control not only contractors but NASA field centers.

Configuration management satisfied the needs of managers, systems engineers, financial experts, and legal personnel. Systems engineers used configuration management to coordinate the designs of the subsystem engineers. Managers found configuration management an ideal lever to control scientists, engineers, and contractors because these groups could no longer make changes without passing through a formal CCB. Financial and legal experts benefited from configuration management because it tied cost and schedule changes to contractual documentation. One business school professor believed that configuration management was influential as a systematic way to resolve group conflicts in NASA projects.[2]

The CCB was the link between the formal hierarchy and the informal working groups. At the board, the project manager and project controller evaluated changes from the standpoint of cost and schedules. Disciplinary representatives evaluated change requests to see if they affected other design areas, while systems engineers determined if the proposed change caused higher-level interactions among components. The CCB ensured frequent communication between the groups. Through the linkage of hierarchies and working groups, and the processes that tied paperwork to hardware, configuration management was the heart of systems management and the key to managerial and military control over scientists and engineers.

Social Gains through Systems Management

Systems management became the standard for space and missile technologies because it promoted the goals of the groups involved. Scientists received credit for conceiving novel technologies. Military officers gained control over radical new weapons and their associated organizations. Engineers earned respect by creating reliable technologies. Managers gained by controlling organizations within a predictable (but hopefully large) budget.

In the 1950s scientists formed an alliance with the military to rapidly create novel weapons. Officers desired quick weapon deployment, while scientists provided the novelty. Within systems management, the role of scientists became standardized. They were to perform systems analyses at the beginning of programs to determine technological feasibility and whether to develop a particular mission or weapon. Scientists also used their quantitative skills to assess the reliability of the new technologies as engineers developed them, acting as credible second sources of information. Managers and military officers frequently used scientists in this capacity, as Ramo-Wooldridge and Aerospace did for the air force against the contractors.

Military officers also gained influence through systems management. Here it is important to distinguish between technical officers and operational officers. The latter led troops into battle and throughout the long history of the military held the reins of power. Technical officers became more important in the 1950s and 1960s. Before the 1950s, air force technical officers lamented their poor career possibilities. With the creation of Air Research and Development Command (ARDC) in 1950 and Air Force Systems Command (AFSC) in 1961, technical officers won for themselves career paths separate from those of their operational brethren. Systems management became the formal set of procedures that allowed them to maintain a military career in technical R&D. Technical officers gained a stable career path and significant power.

In the late 1950s and early 1960s, NASA was run by engineers and for engineers. Although engineers no longer had free rein after that time, NASA remained an engineering organization. So too were ARDC, AFSC, and their laboratories and contractors. Aerospace engineering, and particularly the space program, were the glamour jobs of engineering during this period, with interesting tasks, substantial authority and funding, and excellent opportunities for promotion into managerial or technical positions. Systems management ensured a large role for engineers throughout the design process and kept alive the engineering working groups that allowed engineers to maintain substantial authority. Engineers developed the testing techniques that ensured that their products operated, and hence they guaranteed their own credibility and success. These testing techniques too became part of systems management. Engineers benefited from the creation of systems engineering, which gave the chief systems engineer nearly as much authority as the project manager, who more often than not was also an engineer by training and experience.

Management credibility and authority stem from control of a large organization with funding to match. The power of the purse was the manager's primary weapon, but in the 1950s managers had not yet learned how to use it to control the scientists and engineers. As long as scientists and engineers created novel technologies and funding was plentiful, they could and did claim that they could not predict costs or schedules. Until the technologies reached testing, where failures appeared, managers could not successfully challenge that claim. However, technical failure gave managers the wedge they needed to gain control.

As tasks and projects repeated, managers used past history to predict costs and hold scientists and engineers to estimates based on prior history. Standardizing systems management made costs and schedules more predictable and allowed managers to distinguish between "normal" cost and schedule patterns and "abnormal" patterns that signaled technical or organizational problems. Project managers used this information to control projects and predict outcomes without being completely dependent on technical experts.

Executive managers wanted to know about the current status of projects and about possible new projects. To determine if new projects should be funded, executive managers created "breakpoints" at which they could intervene to continue, modify, or cancel a project. Phased planning implemented these breakpoints and ensured that only limited resources would go toward new projects before executive managers had their say.

From a social viewpoint, each of the four professional groups gained important career niches from the institution of systems management. This helps to explain why it has proven to be a stable method in the aerospace industry. However, this leaves unanswered whether systems management actually made costs predictable or novel products dependable. In the end, none of the social factors would matter if the end products ultimately failed.

The Technical Gains of Systems Management

Technical failures of aerospace projects are hard to hide. Rockets and missiles explode. Satellites stop sending signals back to Earth. Pilots and astronauts die. To the extent that systems management helped prevent these events, it must be deemed a technical success. Systems management methods such as quality assurance, configuration control, and systems integration testing were among the primary factors in the improved dependability of ballistic missiles and spacecraft. Missile reliability in air force and JPL missile programs increased from the 50% range up to the 80% to 95% range, where it remains to this day. JPL's spacecraft programs suffered numerous failures from 1958 to 1963, but after that JPL's record dramatically improved, with a nearly perfect record of success for the next three decades. The manned programs suffered a number of testing failures at the start but had an enviable flight record with astronauts, with the one glaring exception of the *Apollo 204* fire. A strong correlation exists between systems management and reliability improvements.

The nature of reliability argues for the positive influence of systems methods. For aerospace projects to succeed, there must be high-quality components, proper integration of these components, and designed-in backups in case failures occur. Only the last of these is a technology issue in the design sense. The selection and proper integration of components has more to do with rigorous compliance with design and manufacturing standards than it does with new technology. High component quality comes through unflagging attention to manufacturing processes, backed by testing and selection of the best parts. In a nutshell, it is easy to solder a joint or crimp a connector pin but extremely difficult to ensure that workers perform thousands or millions of solders and crimps correctly. Even a worker with the best skills and motivation will make occasional mistakes. In systems management, social processes to rigorously inspect and verify all manufacturing operations ensure high quality across the thousands of workers involved in the process.

Similarly, ensuring proper integration is a matter of making sure that each and every joint is properly soldered, every pin and connector properly crimped, every structure properly handled at all times, and all of these operations rigorously tested. On top of this, "systems testing" checks for design flaws and unexpected interactions among components. In all of these issues, procedures and processes—not new technology—are the keys to success. Systems management provided these rigorous processes and tests.

Once organizations dealt with component problems, they ran into the next most likely cause of failure: interface problems caused by mismatches between designs. By the mid-1960s, both the air force and NASA obsessively concentrated on interface problems, which resulted ultimately from poor communication, poor organization, or both. Engineers and managers recognized that differences in organizational cultures and methods made communication between organizations more difficult than communication within an organization. Miscommunication led to incompatibilities between components and subsystems—incompatibilities often found when components were first connected and tested. More technology was not the solution. Instead, engineers needed improved communication through social processes.

Engineers enforced better communication by creating standard documents and processes. They required that one organization be responsible for analyzing both sides of an interface and that the specifications and analyses be documented in a formal Interface Control Document. Many interface problems were subtler than simple mismatches between physical or electrical components.

For example, engineers at Marshall Space Flight Center found that a "non-liftoff" of a *Mercury-Redstone* test vehicle occurred because the *Mercury* capsule had a different weight than the *Redstone*'s normal warhead, changing the time it took for the launch vehicle to separate from the launch tower. Because the combined launch complex–launch vehicle electronics required that the vehicle lift off at a certain rate, the changed rate led to a shutdown of the launch vehicle as emergency electronics kicked in to abort the launch.

Problems such as these were solvable not through technology but through better engineering communication and better design analysis. Once engineers understood all of the factors, the design solution was usually simple. The problem was making sure the right people had the right information and that someone had responsibility for investigating the entire situation. As ELDO's history shows, getting an organization to pay for a change in an interface was often more difficult than formulating a technical solution. Authority and communication matter most in interface problems and solutions. Better organization and better systems, not better technology, made for reliability in large aerospace projects by standardizing the processes and providing procedures to cross-check and verify each item, from solder joints to astronaut flight procedures. These methods essentially provided insurance for technical success.

How much did this insurance cost? Did systems management lower costs or speed development compared to earlier processes and methods? Concurrency in the 1950s was widely believed to shorten development times, but at enormous cost. The secretary of the air force admitted that the air force could afford only one or two such programs. Schriever contended that concurrency saved money because it shortened development time. Because R&D costs are spent mostly on engineering labor, Schriever argued, shortening development time would reduce labor hours and hence cost. Most other experts then and later disagreed with him. Political scientist Michael Brown contends that concurrency actually led to further schedule slips because problems in one part of the system led to redesigns of other parts, often several times over.[3]

On any given design, having systems management undoubtedly costs more than not having systems management, just as buying insurance costs more than not buying insurance. The real question is whether systems management reduced the number of failures sufficiently, so that it counterbalanced the replacement cost. For example, a 50% rate of reliability for a missile system such as *Atlas* in the late 1950s meant that every other missile failed. With this failure rate, the air force and its contractors could afford to spend up to the cost of an entire second missile in improvements to management processes, if these processes could guarantee success. In other words, at a 50% reliability rate and a cost of $10 million per missile, each *successful* launch costs $20 million. Thus, if process improvements can guarantee success, then spending $10 million or less per missile in management process improvements is cost-effective.

In fact, the early Atlas, Titan, and Corporal projects achieved roughly 40–60% reliability. Reliability improvement programs—that is, systems management processes—improved reliability into the 60–80% range during the 1950s and early 1960s and into the 85–95% range thereafter.[4] The reliability improvement meant that roughly nine out of ten launches succeeded, instead of one out of two. Therefore, systems management could easily have added more than 50% to each missile's cost and still been cost-effective. NASA's efforts to "man-rate" Atlas and Titan could have added 100% to costs for Atlas and Titan and still been cost-effective, because success had to be guaranteed. In fact, considering the potential loss of not only the launchers but also the manned capsules and astronauts, NASA could likely spend 200–500% on launcher improvements and still be cost-effective, considering the low reliability of these vehicles at that time. Pending detailed cost analysis, systems management was probably cost-effective if costs were measured for each successful launch.

Another way to assess systems management is to compare missile and space programs that implemented systems management methods with programs that did not. ELDO provides the most extreme example of little or no systems management. None of its rockets ever succeeded, despite piecemeal introduction of some systems management methods. Comparison of JPL's Ranger program with the contemporary Mariner program provides another example, because the Mariner design was a modification of the Ranger spacecraft. With less systems management, Ranger's first six flights failed, whereas Mariner achieved a 50% success rate, with later Mariner spacecraft performing almost perfectly. After strengthening systems management, Ranger's record was three successes out of four launches.[5] Assuming Ranger and Mariner costs per spacecraft were roughly equal, Mariner cost less per successful flight than early Ranger.

Aside from pure cost considerations, failures hurt an organization's credibility. In the rush to beat the Soviets, early space programs lived the old adage "There is never time to do it right, but there is always time to do it over." They had many failures, but in the early days executive managers were not terribly concerned. By the early 1960s, however, failures mattered; they led to congressional investigations and ruined careers. Systems management responded to the need for better reliability by trying to make sure that engineers "did it right" so they would not have to "do it over."

It is no coincidence that engineers developed systems management for missiles and spacecraft that generally cannot be recovered. When each flight test and each failure means the irretrievable loss of the entire vehicle, thorough planning is much more cost-effective than it is for other technologies that can be tested and returned to the designers. This helps to explain why the bureaucratic methods of systems management work well for space systems but seem too expensive for many Earth-bound technologies. For most technologies, building a few prototypes and performing detailed tests with them before manufacturing is feasible and sensible. Lack of coordination and planning (each of which costs a great deal) can be overcome through prototype testing and redesign of the prototype. This option is not available for most space systems, because they never return.

The evidence suggests that systems management was successful in improving reliability sufficiently to cover the cost per successful vehicle. Although systems management methods were not the only factor involved in these improvements, from the standpoint of reliability, they were critical. Process improvements, not technology improvements, ensured the proper connection and integration of thousands of components. Systems management increased vehicle costs on a per vehicle basis compared to previous methods but reduced costs when reliability is factored in.

Codified Knowledge

In bureaucracies, procedures define the functions of every job and who reports to whom. Procedures specify the communication channels, the data to be communicated, who analyzes data to create information, and who makes decisions based on that information. Systems management defined such procedures for the organization of R&D.

Failure spurred the development of systems management. In the first missile and space projects, few procedures or standards existed, for the simple reason that no one had built space vehicles before. Individuals used methods with which they were familiar, until they found them ineffective. As is typical in pioneering work, individuals made mistakes and did not want to repeat them or have others make the same errors. They developed and documented new processes that avoided their initial errors. Later projects used these methods, sometimes modifying them in accordance with new circumstances.

Procedures served two purposes: to pass on good scientific, engineering, or managerial practices; and to protect the organization and its individuals from external criticism. Organizational and process changes typically followed technical problems rather than preceding them. Documenting the lessons learned from success and failure, and standardizing successful practices, created consistency across the organization. To distinguish this kind of knowledge from other, more familiar forms such as mathematical, scientific, or tacit knowledge, we may call this codified knowledge.[6]

Codified knowledge is knowledge put into verbal or "algorithmic" form, typically documented in explicit written instructions. For example, the military relies upon regulations and procedures because officers frequently rotate to new tasks and positions. The military would degenerate to chaos if it did not have explicit written procedures to document the functions of each position and its processes. Small organizations where each individual understands all tasks need few procedures. However, beyond a certain organizational size, no individual can understand all tasks or processes, and written procedures become more important.

Individuals write procedures to accomplish a specific function. This helps to explain why systems engineers had such a difficult time explaining themselves to their academic colleagues. Engineering researchers in academia defined themselves through a body of theoretical or empirical knowledge. More importantly, they prized the creative process, which cannot follow set rules or strict procedures. Although academic engineers performed specific tasks, such as writing proposals, performing research, and teaching, procedures had little meaning to them, because their research created new knowledge for which no procedures existed. By contrast, systems engineers performed a specific function, and their unique knowledge consisted of algorithm-like processes and procedures to analyze and coordinate the designs of other engineers. Systems engineers needed creativity to solve problems, but the processes that led them to the problems and the methods to coordinate the solutions could be standardized. Hendrick Bode, a systems engineer with Bell Laboratories, compared systems engineers to architects, acting as the bridge between builders and users, designing the overall system, and coordinating the entire effort.[7]

Contrary to the observations of some management theorists, bureaucracy is not inherently antithetical to creativity or to R&D in general. In fact, the history of systems management shows that bureaucracy can be useful to ensure that all parties involved with R&D—the funders, the managers, and the technical experts—have a part in the process. Systems management provides a framework in which R&D flourishes as a stable part of organizations. Furthermore, elements of systems management, such as change control for design coordination, are essential elements of the technical processes in technology creation.

Systematic, Scientific, and Systems Management

Systems engineers and operations researchers often traced the lineage of their techniques to Frederick Taylor's scientific management movement in the early twentieth century. To these mid-twentieth-century spokespeople, Taylor's innovation was the application of scientific methods to the analysis of processes. So too, systems engineering and operations research applied rational thinking to the processes of R&D. Just as systematic management standardized corporate planning and communications at the end of the nineteenth century and scientific management rationalized manufacturing processes in the first decades of the twentieth century, systems management bureaucratized the R&D process in the middle of the twentieth century.[8]

One way to view these three management movements is to identify each with the group over which managers gained authority. In systematic management, executive managers developed methods to coordinate and control lower-level managers and office workers. Scientific management extended their authority over industrial workers. Systems management expanded managerial authority over scientists and engineers. In systematic and scientific management, upper-level managers typically appropriated skills and knowledge from lower-level workers. Systems management did not appropriate skills but rather developed proxy measures to assess R&D workers.

Without the ability to develop technologies themselves, managers and military officers developed schedule and cost measurements to assess progress against a plan. The Program Evaluation and Review Technique (PERT) and the critical path method became popular in management precisely because they gave managers for the first time methods to assess technical work progress without having to rely completely on the technical workers. Configuration management forced technical experts to give the cost and schedule information necessary for managers to develop reasonable plans. Armed with relatively accurate schedule and cost data, managers could then monitor these factors as a proxy for technical progress. Because funding was the primary resource they controlled, managers finally had the means to monitor and control the scientific and engineering teams.

Managers did not eliminate working groups or appropriate the technical knowledge of scientists and engineers. Instead, management superimposed hierarchy over technical teams through the imposition of project management and configuration control. Project management gave one manager control over project funds. Working teams designed the system and periodically reported their progress through design reviews. The products approved in design reviews became baselines, changeable only through management approval tied to cost and schedule. "Management by the numbers" became popular at this time in part because these numbers served as proxies for technical knowledge. In earlier times, managers could directly understand and control their workers. With knowledge workers, this was no longer possible, and "the numbers" became the substitute of choice.

Through systems management, government officials could assess future technologies with systems analysis and control current projects through project management and systems engineering. On highly technical projects such as the military's weapons programs or NASA's space projects, the government's ability to command industry became powerful indeed. Systems analysis, project management, and systems engineering have become standard techniques throughout the government and indeed throughout much of American industry.

People today often criticize NASA for its bureaucratic ways, yet when NASA has relaxed its grip, it has endured failures such as the *Challenger* accident or the *Hubble* telescope's embarrassing optical problems.[9] The "faster, better, cheaper" mantra of the 1990s is of questionable value for space programs. How many of the bureaucratic methods of systems management can be eliminated before there are large-scale failures? Recent events, such as the loss of NASA's 1999 Mars probes, show that NASA cannot eliminate many systems management methods. It is folly to think that it could be otherwise for space projects.

Proponents of the "faster, better, cheaper" approach want to return NASA to the days of frequent, relatively inexpensive spacecraft and launchers built by small, informal teams. As we have seen, however, the "good old days" were also times of frequent failures, huge cost overruns, and common schedule slips. NASA managers and engineers created the formal mechanisms of systems management explicitly in response to the problems of "ad-hocracy." Systems management, as recent spacecraft failures once again prove, is cost-effective on a per-successful-flight basis for space projects.

Another criticism comes by comparison with Japan. The Japanese developed a managerial system based on quality. This system, according to promoters of methods such as total quality management (TQM), is superior to the American system because it focuses on the "voice of the customer" and because workers on the factory floor have a voice in improving the manufacturing process. The Japanese method is a direct outgrowth of American-style Taylorism and has often been roundly criticized in Japan as being too rigid and controlling. The quality methods developed in Japan stem from the same concerns with reliability that the American aerospace industry had. Whereas Japanese managers and engineers focused on the reliability of mass production processes, their American counterparts concentrated on the reliability of R&D. Both approaches used inputs from the working teams, in the United States with R&D scientists and engineers, and in Japan with the manufacturing engineers and production line workers. The TQM promoters believe that paying more attention to "the customer" and to quality will solve American ills.[10]

The Japanese approach, bred in the "stable-tech" automotive and similar mass production industries, may well be suitable for American mass production industries but is not well suited to R&D. Japanese industry has been quite successful at copying American innovations and mass-producing them but far less successful at producing its own innovations. The Japanese are far more likely to gain from American systems management for their R&D than Americans are to gain from applying manufacturing-based TQM to R&D, for the simple reason that systems management developed as an R&D process and TQM did not.

One last consideration points to the broader importance of systems management. Systems management methods have spread far beyond aerospace. As early as 1972, Ida Hoos described numerous government organizations that adopted systems analysis and probably other elements of systems management as well. She decried this as the unwarranted spread of technocratic methods to areas in which they were inappropriate. That is undoubtedly true.[11]

One significant place where systems management spread is the computer software industry. The information industry is now the current exemplar for American industrial dominance. Computer software has supplanted hardware as the glamour product. Yet software development is exactly where systems management methods developed in the computer industry. In the late 1950s, the largest software company was the System Development Corporation (SDC), which spun out of the RAND Corporation to develop software and training programs for air force air-defense systems. Programmers trained in systems methodology at SDC on air force programs diffused throughout the United States, spreading the systems approach for programming far and wide.[12]

More than with any hardware artifact, software development is a pure process, and the final product is itself a process. Current methodologies in software development, such as structured or object-oriented programming, are variations on systems themes: simplifying interfaces, enhancing communications, considering the entire product, and dividing tasks into simple work packages. Whatever might be said about software, the industry is rarely deemed overconservative or lacking in innovation. Perhaps systems approaches prevent computer programmers and computer scientists from finding a radically better way of developing software. Or perhaps they are the only thing preventing software from sinking into total chaos.

The officers, managers, engineers, and scientists who created systems management in the first two decades of the Cold War did so because they believed in the efficacy of technology to protect and promote the values of the United States. After this time, the apparent effectiveness of these methods in creating missile, space, and computing technologies led technologists and managers in other nations to mimic Americans. Through the combined efforts of these groups of people, technological innovation has become a standardized commodity throughout the Western world. Systems management has tamed R&D.

Political, military, and business leaders now plan for new technologies years in advance, using the services of scientists and engineers to promote their visions of the future. Systems management has become one of the primary tools of our technological civilization. Change is now the norm, a standardized product of systems management.

Notes

Introduction

James R. Beniger, *The Control Revolution: Technological and Economics Origins of the Information Society* (Cambridge: Harvard University Press, 1986).

1. Alfred D. Chandler Jr., *The Visible Hand: The Managerial Revolution in American Business* (Cambridge: Belknap Press of the Harvard University Press, 1977).

2. JoAnne Yates, *Control Through Communication: The Rise of System in American Management* (Baltimore: Johns Hopkins University Press, 1989).

3. Robert Kanigel, *The One Best Way: Frederick Winslow Taylor and the Enigma of Efficiency* (New York: Penguin Putnam, 1998).

4. Olivier Zunz, *Making America Corporate: 1870–1920* (Chicago: University of Chicago Press, 1990). Hugh G. H. Aitken, *Taylorism at Watertown Arsenal: Scientific Management in Action, 1908–1915* (Cambridge: Harvard University Press, 1960).

5. Management texts of the time confirm this lack of attention. A typical example is Ralph Currier Davis, *Industrial Organization and Management,* 3d ed. (New York: Harper & Brothers, 1957).

6. Stephen Waring, "Peter Drucker, MBO, and the Corporatist Critique of Scientific Management," in Daniel Nelson, ed., *A Mental Revolution: Scientific Management since Taylor* (Columbus: Ohio State University Press, 1992), 205–36.

7. Max Weber, *The Theory of Social and Economic Organization* (New York: Oxford University Press, 1947). For the classic analysis of Weber's theory, see Reinhard Bendix, *Max Weber, An Intellectual Portrait* (New York: Doubleday, 1960).

8. R. Cowan and D. Foray, "The Economics of Codification and the Diffusion of Knowledge," *Industrial and Corporate Change* 6, no. 3 (1997): 595–622. Naomi R. Lamoreaux and Daniel M. G. Raff, eds., *Coordination and Information: Historical Perspectives on the Organization of Enterprise* (Chicago: University of Chicago Press, 1995). Laurence Prusak, ed., *Knowledge in Organizations* (Boston: Butterworth-Heinemann, 1997).

9. Beniger, *The Control Revolution.* Ross Thomson, *Learning and Technological Change* (New York: St. Martin's Press, 1993).

10. Wiebe E. Bijker, Thomas Hughes, and Trevor Pinch, eds., *The Social Construction of Technological Systems* (Cambridge: MIT Press, 1987). Donald MacKenzie and

Judy Wachmann, eds., *The Social Shaping of Technology* (Philadelphia: Open University Press, 1999).

11. Stephen B. Johnson, *The United States Air Force and the Culture of Innovation, 1945–1965* (Washington, D.C.: USAF History Support Office, 2002).

12. Some of the finer examples include the following: Arnold S. Levine, *Managing NASA in the Apollo Era,* SP-4102 (Washington, D.C.: NASA, 1982). Diane Vaughan, *The Challenger Launch Decision* (Chicago: University of Chicago Press, 1996). Joan Lisa Bromberg, *NASA and the Space Industry* (Baltimore: Johns Hopkins University Press, 1999). Andrew J. Dunar and Stephen P. Waring, *Power to Explore: A History of Marshall Space Flight Center 1960–1990,* SP-4313 (Washington, D.C.: NASA, 2000). Howard McCurdy, *Inside NASA* (Baltimore: Johns Hopkins University Press, 1993). W. Henry Lambright, *Powering Apollo: James E. Webb of NASA* (Baltimore: Johns Hopkins University Press, 1995). Phillip K. Tompkins, *Organizational Communication Imperatives: Lessons of the Space Program* (Los Angeles: Roxbury, 1992). John Lonnquest, "The Face of Atlas: General Bernard Schriever and the Development of the Atlas Intercontinental Ballistic Missile 1953–1960" (Ph.D. diss., Duke University, 1996). Clayton R. Koppes, *JPL and the American Space Program: A History of the Jet Propulsion Laboratory* (New Haven: Yale University Press, 1982). Harvey M. Sapolsky, *The Polaris System Development: Bureaucratic and Programmatic Success in Government* (Cambridge: Harvard University Press, 1972). Irving Brinton Holley Jr., *Buying Aircraft: Materiel Procurement for the Army Air Forces,* vol. 7 of Stetson Conn, ed., *United States Army in World War II* (Washington, D.C.: Dept. of the Army, 1964). R. Cargill Hall, *Lunar Impact: A History of Project Ranger,* SP-4210 (Washington, D.C.: NASA, 1977).

One | Social and Technical Issues of Spaceflight

Jean-Jacques Servan-Schreiber, *The American Challenge,* trans. Robert Steel (New York: Avon Books, 1969), 168 (emphasis in original). First appeared as *Le Défi American* (Paris: Editions Denoel, 1967).

1. *Apollo Program Management,* Staff Study for the Subcommittee on NASA Oversight of the Committee on Science and Astronautics, U.S. House of Representatives, 91st Congress, 1st Session, July 1969 (Washington, D.C.: Government Printing Office, 1969).

2. Robert C. Seamans Jr. and Frederick I. Ordway, "The Apollo Tradition: An Object Lesson for the Management of Large-Scale Technological Endeavors," in Frank Davidson and C. Lawrence Meador, eds., *Macro-Engineering and the Future: A Management Perspective* (Boulder, Colo.: Westview Press, 1982), 11.

3. "Long Shadow from Apollo," *Nature* 226 (1970): 197.

4. J. J. Beattie and J. de la Cruz, "ESRO and the European Space Industry," *ESRO Bulletin* 3 (1967): 3–8.

5. Michael J. Neufeld, *The Rocket and the Reich: Peenemünde and the Coming of the Ballistic Missile Era* (New York: Free Press, 1995). Milton Lomask, *Robert Goddard:*

Space Pioneer (Champaign, Ill.: Garrard, 1972). J. D. Hunley, "A Question of Antecedents: Peenemünde, JPL, and the Launching of U.S. Rocketry," in Roger D. Launius, ed., *Organizing for the Use of Space: Historical Perspectives on a Persistent Issue,* AAS History Series, vol. 18, R. Cargill Hall, series ed. (San Diego: Univelt for the AAS, 1995), 1–31. Wernher von Braun and Frederick Ordway III, *History of Rocketry and Space Travel* (New York: Crowell, 1966). William E. Burrows, *This New Ocean: The Story of the First Space Age* (New York: The Modern Library, 1998). T. A. Heppenheimer, *Countdown: A History of Space Flight* (New York: Wiley, 1997).

6. A standard introductory text is George Paul Sutton, *Rocket Propulsion Elements: An Introduction to the Engineering of Rockets,* 6th ed. (New York: Wiley, 1992).

7. Roger E. Bilstein, *Stages to Saturn: A Technological History of the Apollo/Saturn Launch Vehicles,* SP-4206 (Washington, D.C.: NASA, 1980), chap. 4.

8. Missile designers virtually monopolized early symposia and conferences on reliability and quality assurance, which were sponsored by the military services. For example, see *Proceedings, National Symposium on Quality Control and Reliability in Electronics,* New York City, November 12–13, 1954, sponsored by the Professional Group on Quality Control, Institute of Radio Engineers, and the Electronics Technical Committee, American Society for Quality Control (New York: Institute of Radio Engineers, 1955).

9. See, for example, *Bimonthly Summary Report No. 36a, Supplement to Combined Bimonthly Summary No. 36, the Corporal Guided Missile XSSM-A-17,* 1 August 1953, JPLA; and *Bimonthly Summary Report No. 37a, Supplement to Combined Bimonthly Summary No. 37, the Corporal Guided Missile XSSM-A-17,* 1 October 1953, JPLA.

10. Wilfrid J. Mayo-Wells, "The Origins of Space Telemetry," in Eugene M. Emme, ed., *The History of Rocket Technology* (Detroit: Wayne State University Press, 1964), 253–70.

11. "Report on Launch of F6/2 Vehicle," Note by the Secretariat, ELDO/T(68)1, Neuilly, 8 February 1968, HAEUI ELDO Fond 2992.

12. For example, see "Report on Launch of F6/1 Vehicle," Note by the Secretariat, ELDO/T(67)24, Paris, 22 September 1967, HAEUI ELDO Fond 2977.

13. The following description of satellite problems comes from the author's experience at Martin Marietta Corporation in the 1980s.

14. See Charles Perrow, *Normal Accidents* (New York: Basic Books, 1984).

15. On Ranger 6, see R. Cargill Hall, *Lunar Impact: A History of Project Ranger,* SP-4210 (Washington, D.C.: NASA, 1977), 258–61. On the Europa problem, see "Comments by the Secretariat on Interface Problems Following the First Report of the Corps of Inspectors," Note by the Secretariat, ELDO/T(67)5, Paris, 7 February 1967, HAEUI Fond 9458.

16. "Draft Resolution-Integrating Group," ELDO/C(66)WP/16, Paris, 29 September 1966, ELDO Fond 1229, HAEUI. L. T. D. Williams, Chairman, Corps of Inspectors, letter to Secretary-General, ELDO/CECLES, ELDO/T(67)5, London, 17 December 1966, HAEUI ELDO Fond 2958. *Organisation Européenne pour la mise au point et la construction de lanceurs d'engins spatiaux, structure et table des matières du rapport*

de synthèse de la Commission de Revue de Projet, 1ère partie, Rapport de Synthèse de la Commission de Revue de Projet, Neuilly, 19 Avril 1972, ELDO/CRP(72)38, HAEUI Fond 2885.

17. Arnold Frutkin, letter to Professor Pierre Auger, Director-General, ESRO, 10 July 1964, HAEUI ESRO Fond 50151. *Project Mercury Status Report No. 2 for Period Ending April 30, 1959,* NASA, Langley STG, NASAHO.

18. Stuart Leslie, *The Cold War and American Science* (New York: Columbia University Press, 1993). Roger L. Geiger, *Research and Relevant Knowledge: American Universities since World War II* (New York: Oxford University Press, 1993). Daniel J. Kevles, *The Physicists: The History of a Scientific Community in Modern America* (Cambridge: Harvard University Press, 1971).

19. Michael E. Brown, *Flying Blind: The Politics of the U.S. Strategic Bomber Program* (Ithaca, N.Y.: Cornell University Press, 1992). John Lonnquest, "The Face of Atlas: General Bernard Schriever and the Development of the Atlas Intercontinental Ballistic Missile 1953–1960" (Ph.D. diss., Duke University, 1996).

20. Norman Friedman, *The Fifty Year War: Conflict and Strategy in the Cold War* (Annapolis, Md.: Naval Institute Press, 2000).

21. House Committee on Armed Services, *Hearings, Weapons System Management and Team System Concept in Government Contracting,* 86th Congress, 1st Session, 1959. House Committee on Government Operations, Subcommittee on Military Operations, *Organization and Management of Missile Programs,* 86th Congress, 1st Session, Report No. 1121, 1959.

Two | Creating Concurrency

Colonel Norair M. Lulejian, Deputy for Technology, Space Systems Division, "Scheduling Invention," in *AFSC Management Conference,* Monterey, Calif., 2–4 May 1962 (Washington, D.C.: Air Force Systems Command, 1962), 1-4-1.

1. This chapter includes materials from chapters 2 and 3 of Stephen B. Johnson, *The United States Air Force and the Culture of Innovation, 1945–1965* (Washington, D.C.: USAF History Support Office, 2002).

2. See an extract of the Signal Corps–Wright Brothers contract in G. Van Reeth, "Le contrat avec intéressement comme instrument de gestion," *ESRO/ELDO Bulletin* 7 (October 1969): 5.

3. Roger E. Bilstein, *Flight in America,* rev. ed. (Baltimore: Johns Hopkins University Press, 1994), 70–72. Alex Roland, *Model Research: The National Advisory Committee for Aeronautics 1915–1958,* SP-4103 (Washington, D.C.: NASA, 1985), 73–119. I. B. Holley Jr., *Ideas and Weapons* (New York: Yale University Press, 1953).

4. Jacob Vander Meulen, *The Politics of Aircraft* (Lawrence, Kan.: University Press of Kansas, 1991). Allen Kaufman, *In the Procurement Officer We Trust: Constitutional Norms, Air Force Procurement and Industrial Organization, 1938–1948* (Cambridge: MIT Defense and Arms Control Studies Program Working Paper, January 1996), 21–29.

5. Benjamin N. Bellis, L/Col USAF Office DCS/Systems, "The Requirements for Configuration Management During Concurrency," in *AFSC Management Conference,* Monterey, Calif., 2–4 May 1962 (Washington, D.C.: Air Force Systems Command, 1962), 2-24-1–2-24-4. "Combat Ready Aircraft," an Air Force Study prepared by Deputy Chief of Staff, Development, April 1951, Marvin C. Demler Papers, AFHRA 168.7265-236, 16–18.

6. Robert L. Perry, *System Development Strategies: A Comparative Study of Doctrine, Technology, and Organization in the USAF Ballistic and Cruise Missile Programs, 1950–1960,* RAND Corporation Memorandum RM-4853-PR (Santa Monica, Calif.: RAND Corporation, August 1966), 14–15. Lt. General Donald L. Putt, Oral History Program Interview with James C. Hasdorff, 1–3 April 1974, AFHRA K239.0512-724, 23–24, 123.

7. Irving Brinton Hollcy Jr., *Buying Aircraft: Materiel Procurement for the Army Air Forces,* vol. 7 of Stetson Conn, ed., *United States Army in World War II* (Washington, D.C.: Dept. of the Army, 1964), chaps. 13, 15, 16.

8. Kaufman, *In the Procurement Officer,* 49–55.

9. Von Kármán, "Where We Stand," 22 August 1945, and "Science: The Key to Air Supremacy," 15 December 1945, in Michael H. Gorn, ed., *Prophecy Fulfilled: "Toward New Horizons" and Its Legacy* (Washington, D.C.: Air Force History and Museums Program, 1994), 17–82 and 83–190. Dik A. Daso, Maj., USAF, *Architects of American Air Supremacy: Gen Hap Arnold and Dr. Theodore von Kármán* (Maxwell Air Force Base, Ala.: Air University Press, 1997), 58–61.

10. Thomas A. Sturm, *The USAF Scientific Advisory Board: Its First Twenty Years 1944–1964* (Washington, D.C.: USAF Historical Division Liaison Office, 1967). Donald Ralph Baucom, "Air Force Images of Research and Development and Their Reflections in Organizational Structure and Management Policies" (Ph.D diss., University of Oklahoma, 1976), 11–17, 44–51.

11. John Lonnquest, "The Face of Atlas: General Bernard Schriever and the Development of the Atlas Intercontinental Ballistic Missile 1953–1960" (Ph.D. diss., Duke University, 1996), 57–59. Jacob Neufeld, "Bernard A. Schriever: Challenging the Unknown," in John L. Frisbee, ed., *Makers of the United States Air Force* (Washington, D.C.: Office of Air Force History, 1987), 281–84. Bernard A. Schriever, telephone interview by author, 4 March 1999, Grand Forks, N.D., tape recording at USAF HQ HSO.

12. Robert Sigethy, "The Air Force Organization for Basic Research 1945–1970: A Study in Change" (Ph.D. diss., American University, 1980), 25–26. Nick A. Komons, *Science and the Air Force: A History of the Air Force Office of Scientific Research* (Arlington, Va.: Office of Aerospace Research, 1966), 10–11.

13. Michael H. Gorn, *Harnessing the Genie: Science and Technology Forecasting for the Air Force 1944–1986* (Washington, D.C.: Office of Air Force History, 1988), 48.

14. Putt, interview by Hasdorff, 79.

15. Ibid., 79–82.

16. Ibid.

17. *Research and Development in the United States Air Force* [Ridenour Report],

Scientific Advisory Board, 21 September 1949, AFHRA 168.1511-1. Edmund Beard, *Developing the ICBM: A Study in Bureaucratic Politics* (New York: Columbia University Press, 1976), 107–17. Michael H. Gorn, *Vulcan's Forge: The Making of an Air Force Command for Weapons Acquisition (1950–1985)* (Washington, D.C.: Office of History, Air Force Systems Command, 1989), 7–19. Dennis J. Stanley and John J. Weaver, *An Air Force Command for R&D, 1949–1976: The History of ARDC/AFSC* (Washington, D.C.: Office of History, Headquarters, Air Force Systems Command, 1976).

18. Bruce L. R. Smith, *The RAND Corporation: Case Study of a Nonprofit Advisory Committee* (Cambridge: Harvard University Press, 1966). David R. Jardini, "Out of the Blue Yonder: The RAND Corporation's Diversification into Social Welfare Research, 1946–1968" (Ph.D. diss., Carnegie Mellon University, 1996).

19. Allan A. Needell, *Science, Cold War, and the American State* (Amsterdam: Harwood, 2000), chap. 5.

20. Lonnquest, "Face of Atlas," 213–14.

21. Jacob Vander Muelen, *Building the B-29* (Washington, D.C.: Smithsonian Institution Press, 1995), 11–29. Perry, *System Development Strategies,* 15–16.

22. Vincent C. Jones, *Manhattan: The Army and the Atomic Bomb* (Washington, D.C.: Center of Military History, United States Army, 1985). Richard Rhodes, *The Making of the Atomic Bomb* (New York: Touchstone, Simon & Schuster, 1986), 486–96.

23. Rhodes, *Atomic Bomb,* 454–55, 460. Lillian Hoddeson et al., *Critical Assembly: A Technical History of Los Alamos during the Oppenheimer Years, 1943–1945* (Cambridge: Cambridge University Press, 1993), 1–3, chap. 8.

24. Von Kármán, "Where We Stand," 37.

25. Ridenour Report, IX-1-2.

26. Perry, *System Development Strategies,* 14–17.

27. Colonel M. C. Demler, Presentation to Chief of Staff and Comptroller, WADC, on Management Practices, 15 February 1952, AFHRA 168.7265-235, 4.

28. "Combat Ready Aircraft," 2.

29. "Combat Ready Aircraft," 9–14.

30. "Combat Ready Aircraft," 15.

31. Ridenour Report, IX-2.

32. "Combat Ready Aircraft," 19–24. Perry, *System Development Strategies,* 18–20.

33. "Organizational Philosophy of the Air Force," presented by Maj. General Donald L. Putt, Commanding General, WADC, to Chiefs of WADC Laboratories, 2 April 1952, AFHRA 168.7265-235, 6–8.

34. Robert J. Reed, "New AF Policy Means More Competition—More Selling," *Aviation Age* 19, no. 8 (August 1953): 21–22 (emphasis in original).

35. Gorn, *Vulcan's Forge,* 19–20.

36. AFR 20-10, "Weapons System Project Office," 16 October 1951. Gorn, *Vulcan's Forge,* 28–29.

37. The Cook-Craigie Group included Harvard Business School professor Edmund Learned (who had done organizational studies for the air force headquarters

before), Bell Laboratories chief Mervin Kelly, Brig. Gen. Floyd Wood of ARDC, Edward Wells of Boeing, and Col. William E. Sault of AMC. See Gorn, *Vulcan's Forge,* 27.

38. Once boosted through the atmosphere, ballistic missiles simply fall to their target using gravity, although the reentry vehicles that carry the warheads are often guided once they enter the atmosphere. Guided missiles require active guidance to their targets through the atmosphere.

39. Beard, *Developing the ICBM,* 16–20, 120–28.

40. Lonnquest, "Face of Atlas," 82–97. Schriever, interview with author, 4 March 1999.

41. Lonnquest, "Face of Atlas," 98–112. Simon Ramo, telephone interview with author, 7 December 1999, Grand Forks, N.D., notes with author.

42. Ramo and Wooldridge had just left eccentric billionaire Howard Hughes's company to form one of their own. Jacob Neufeld, *Ballistic Missiles in the United States Air Force 1945–1960* (Washington, D.C.: Office of Air Force History, United States Air Force, 1990), 98–99, and Donald McKenzie, *Inventing Accuracy: A Historical Sociology of Nuclear Missile Guidance* (Cambridge: MIT Press, 1990), 109–10. Lonnquest, "Face of Atlas," 113–16. Ramo, interview with author, 7 December 1999.

43. "Recommendations of the Tea Pot Committee," 1 February 1954, in Neufeld, *Ballistic Missiles,* 260–61.

44. Neufeld, *Ballistic Missiles,* 106. *Space and Missile Systems Organization: A Chronology, 1954–1979,* Office of History, Headquarters, Space Division, AFSC Historical Publication, n.d., Space and Missile Center History Office. Bernard A. Schriever, telephone interview with author, 25 March 1999, Grand Forks, N.D., tape recording at USAF HQ HSO. The air force initially wanted to elevate Maj. General James McCormack to this position, but health problems intervened.

45. Schriever believed that ICBMs were significantly more complex because of the heterogeneity and immaturity of the technologies in their component parts.

46. "In-house" means building components inside the company instead of purchasing them from outside. Bernard A. Schriever, telephone interview with author, 13 April 1999, Grand Forks, N.D., tape recording at USAF HQ HSO.

47. House Committee on Government Operations, *Organization and Management of Missile Programs,* 86th Congress, 1st Session, House Report No. 1121, 1959, 75.

48. Schriever, interview with author, 25 March 1999.

49. Lonnquest, "Face of Atlas," 163–70. Beard, *Developing the ICBM,* 172–78. Jacob Neufeld, ed., *Reflections on Research and Development in the United States Air Force: An Interview with General Bernard A. Schriever and Generals Samuel C. Phillips, Robert T. Marsh, and James H. Doolittle, and Dr. Ivan A. Getting,* conducted by Dr. Richard Kohn (Washington, D.C.: Center for Air Force History, 1993), 62–63. Assistant Secretary of Defense Donald Quarles, formerly head of Bell Laboratories, insisted that R-W be placed as "line," not "staff," in Schriever's organization. That is, R-W would not merely be consultants but would have directive authority. This no doubt reflected his convictions about the technical people having authority, as they did at Bell Labs.

Ivan A. Getting, *All in a Lifetime: Science in the Defense of Democracy* (New York: Vantage, 1989), 385–86.

50. Neufeld, *Ballistic Missiles,* 114. Davis Dyer, *TRW: Pioneering Technology and Innovation since 1900* (Boston: Harvard Business School Press, 1998), 180–81. Ramo, interview with author, 7 December 1999.

51. Lonnquest, "Face of Atlas," 262–63.

52. *Space and Missile Systems Organization,* 18–19, 23. Ethel M. DeHaven, *Aerospace: The Evolution of USAF Weapons Acquisition Policy 1945–1961* (Los Angeles, Calif.: Deputy Commander for Aerospace Systems Historical Office, August 1962), 34. Quotation from Memorandum for the record from General Schriever, Subject: Budget strategy, 2 October 1954, AFHRA Microfilm 35257. Memorandum to Colonel Sheppard and Colonel Terhune from General Schriever, Subject: FY 1955 reprogramming actions, 8 October 1954, AFHRA Microfilm 35257.

53. Memorandum to General Power from General Schriever, Subject: Policy and procedure for R-W study contracts, 2 October 1954, AFHRA Microfilm 35257. Memorandum to Colonel Sheppard from General Schriever, Subject: Handling of contractors, 26 October 1954, AFHRA Microfilm 35257. Memorandum to Colonel Terhune from General Schriever, Subject: Competitions for development hardware contracts, 26 October 1954, AFHRA Microfilm 35257. Memorandum to Colonel Terhune from General Schriever, Subject: R-W participation on evaluation boards, 24 January 1955, AFHRA Microfilm 35257. The evidence for R-W participation in Source Selection Boards is contradictory but points to R-W contributing information but the air force making final decisions with R-W personnel not present.

54. Letter to All Centers from Lt. General Thomas Power, 24 February 1955, Subject: Eligibility of the Ramo-Wooldridge Corporation and Thompson Products, Inc. to bid, AFHRA Microfilm 35258, 168.7171-80. Memorandum to Colonel Ford from General Schriever, Subject: General Dynamics-Convair propaganda campaign, 18 March 1955, AFHRA Microfilm 35257.

55. Rough draft of "The Atlas Nose Cone," 12 December 1956, AFHRA Microfilm 35267.

56. *Space and Missile Systems Organization,* 26–27. Neufeld, *Ballistic Missiles,* 119–30.

57. Memorandum for Colonel Sheppard, from Brig. Gen. B. A. Schriever, Subject: Money, 11 January 1955, AFHRA Microfilm 35257. Quotation from Memorandum for General Schriever, from William A. Sheppard, Deputy Commander, Program Management, Subject: Money, 12 January 1955, AFHRA Microfilm 35257.

58. Neufeld, *Ballistic Missiles,* 138–39. Lonnquest, "Face of Atlas," 224–44. Memorandum to Colonel Sheppard from General Schriever, Subject: Subject presentation, 1 April 1955, AFHRA Microfilm 35257.

59. Lonnquest, "Face of Atlas," 245–48. Quotation from General Bernard A. Schriever, Oral History Interview with Dr. Edgar F. Puryear Jr., 15 and 29 June 1977, AFHRA K239.0512-1492, 3. Neufeld, ed., *Reflections,* 56–57. *Space and Missile Systems*

Organization, 28–29. Memorandum to Mr. Trevor Gardner from General Schriever, Subject: Meeting on 6 September 1955 to discuss management fund for ICBM program, 6 September 1955, AFHRA Microfilm 35258, 168.7171-80. DeHaven, *Aerospace,* 43–44. In fact, two groups formed to investigate the management situation. Deputy Secretary of Defense Reuben Roberston led this second group. It generated recommendations similar to those of the Gillette Committee.

60. Schriever, interview with Puryear, 6.

61. Neufeld, *Ballistic Missiles,* 136–37. Gillette Report, 10 November 1955, in Neufeld, *Ballistic Missiles,* 279; *Organization and Management of Missile Programs,* 11th Report by the Committee on Government Operations, 86th Congress, 1st Session, House Report No. 1121 (Washington, D.C.: Government Printing Office, 1959), 23–29. DeHaven, *Aerospace,* 53–54.

62. "Notes Transcribed from an Interview Conversation with Colonel Ray E. Soper, Vice Commander BSD, on the Final Day of His Service with USAF before Retirement," BSD (BEH) NAFB, Cal 92409, 29 November 1966, AFHRA Microfilm 30015, 3.

63. Perry, *System Development Strategies,* 71.

64. Lonnquest, "Face of Atlas," 300–302. Harvey Sapolsky found that on the navy's Polaris program, the use of new managerial methods helped promote an aura of competence, shielding the program from criticism. This is similar to what happened for Schriever on the ICBM program. See Harvey M. Sapolsky, *The Polaris System Development: Bureaucratic and Programmatic Success in Government* (Cambridge: Harvard University Press, 1972).

65. Lonnquest, "Face of Atlas," 300–311; quotations at 311.

66. Ethel M. DeHaven, *Air Materiel Command Participation in the Air Force Missiles Program through December 1957* (Inglewood, Calif.: Office of Information Services, Air Materiel Command Ballistic Missile Office, 1958), 177–81. *Organization and Management of Missile Programs,* 30–32. Memorandum to Colonel Terhune from General Schriever, Subject: R-W participation on evaluation boards, 24 January 1955, AFHRA Microfilm 35257.

67. Memorandum to Colonel Terhune from General Schriever, Subject: Procedures governing development of facilities, 17 March 1955, AFHRA Microfilm 35257. DeHaven, *Air Materiel Command Participation,* 177–81. Perry, *System Development Strategies,* 74, 79.

68. DeHaven, *Air Materiel Command Participation,* 290–91.

69. Memorandum to General Schriever from Colonel Terhune, Subject: Continuous monitoring of USAF Atlas development contracts, 21 December 1954, AFHRA Microfilm 35257. Memorandum to General Schriever from Colonel Sheppard, Subject: (unclassified) WDD fiscal management with chart, 24 January 1955, Item 108, AFHRA Microfilm 35257. Memorandum to staff from William A. Sheppard, the Deputy Commander for Program Management, Subject: Contractor supervision, 17 January 1955, AFHRA Microfilm 35257.

70. DeHaven, *Air Materiel Command Participation,* 171.
71. Ibid., 182–83.
72. Ibid., 183.

Three | From Concurrency to Systems Management

Benjamin N. Bellis, L/Col USAF Office DCS/Systems, "The Requirements for Configuration Management During Concurrency," in AFSC Management Conference, Air Force Systems Command, Andrews Air Force Base, Washington, D.C., AFHRA Microfilm 26254, 5-24-4 (emphasis in original).

1. This chapter contains materials from chapters 3 and 5 of Stephen B. Johnson, *The United States Air Force and the Culture of Innovation, 1945–1965* (Washington, D.C.: USAF History Support Office, 2002).

2. I will not describe here the development of systems engineering's roots in the computer industry, which also involved interaction with Bell Labs and MIT. See Johnson, *United States Air Force and the Culture of Innovation.* Stephen B. Johnson, "Three Approaches to Big Technology: Operations Research, Systems Engineering, and Project Management," *Technology and Culture* 38, no. 4 (1997): 891–919.

3. Ivan A. Getting, telephone interview with author, 30 October and 6 November 1998, Grand Forks, N.D., tape recording at USAF HQ HSO. Dean Wooldridge, one of the two founders of Ramo-Wooldridge, believed also that systems engineering originated at Bell Labs and was "working its way into the industry" by the early 1950s. Davis Dyer, *TRW: Pioneering Technology and Innovation since 1900* (Boston: Harvard Business School Press, 1998), 443, chap. 7, n. 43.

4. David A. Mindell, "Automation's Finest Hour: Radar and System Integration in World War II," paper presented to Symposium on the Spread of the Systems Approach, Dibner Institute, Cambridge, Mass., 5 May 1996, 6–9. Ivan A. Getting, *All in a Lifetime: Science in the Defense of Democracy* (New York: Vantage, 1989), 117–28.

5. "Statement of Relationships between the Bureau of Ordnance, U.S. Navy and the National Defense Research Committee, Office of Science Research and Development, on the Development and Production of the Gunfire Control System Mark 56," reprinted in Getting, *All in a Lifetime,* 186.

6. Mindell, "Automation's Finest Hour," 9.

7. Johnson, "Three Approaches to Big Technology," 914.

8. *Research and Development in the United States Air Force* [Ridenour Report], Scientific Advisory Board, 21 September 1949, AFHRA 168.1511-1, IX-1, 2.

9. "Organizational Philosophy of the Air Force," presented by Maj. General Donald L. Putt, Commanding General, WADC, to Chiefs of WADC Laboratories, 2 April 1952, AFHRA 168.7265-235.

10. Dyer, *TRW,* 168–70.

11. Simon Ramo, telephone interview with author, 24 August 2000, Colorado Springs, Colo., notes with author.

12. Simon Ramo, telephone interview with author, 7 December 1999, Grand Forks,

N.D., notes with author. Gardner and Ramo were good friends going back to the 1940s, when both lived in the same apartment complex while working for General Electric.

13. Charles Terhune, telephone interview with author, 24 and 30 September 1998, tape recording at USAF HQ HSO. Memorandum to General Power from General Schriever, Subject: Ramo-Wooldridge contract administration and study subcontracts, 1 November 1954, AFHRA Microfilm 35257.

14. Memorandum to Dr. Ramo from General Schriever, Subject: R-W organization and procedures, 20 April 1955, AFHRA Microfilm 35257.

15. Letter with attached charts to General Schriever and others from Simon Ramo, 26 April 1955, Subject: The Ramo-Wooldridge Corporation's Guided Missile Research Division Organization, AFHRA Microfilm 35258, 168.7171-80.

16. Ibid.

17. *Weapons System Management and Team System Concept in Government Contracting*, Hearings before the Subcommittee for Special Investigations of the Committee on Armed Services, House of Representatives, 86th Congress, 1st Session (Washington, D.C.: Government Printing Office, 1959), 202–4. Quotation from Colonel Edward N. Hall, Oral History Interview with Jack Neufeld, 11 July 1989, AFHRA K239.0512-1820, 6. TAB A, RW-EO(I)-011-56-? (unreadable), 4th rough draft, 27 December 1956, "The Atlas Guidance System," AFHRA Microfilm 35267. TAB C, RW-EO(I)-011-56-C, 3d rough draft, 28 December 1956, "Ramo-Wooldridge's Contribution to the Titan Missile," AFHRA Microfilm 35267. TAB D, RW-EO(I)-011-56-D, 17 January 1957, "The IRBM Preliminary Analysis and Design Specification," AFHRA Microfilm 35267.

18. "Excerpt from Supplement Agreement #9," 24 October 1956, AF18(600)-1190, AFHRA Microfilm 35267.

19. Memorandum for the record from Lt. Colonel Bogert, Subject: Information submitted to Colonel Covemar in General Irvine's office regarding R-W, 26 October 1956, AFHRA Microfilm 35267. Memorandum to CMDR/AFMTC from General Schriever, Subject: Management requirements (with attachment—draft of integration of WDD into normal structure of ARDC), 6 November 1956, AFHRA Microfilm 35267.

20. Some scholars have questioned whether the control room served any real purpose aside from showmanship. My conclusion is that some of the people in the WDD saw that it had a "show" purpose and concluded that this was its only purpose. After interviewing Charles Terhune, I am convinced that it served a real purpose in giving the "official" status at any time and in providing a central information repository. Terhune, telephone interview with author, 14 and 20 October 1998, tape recording at USAF HQ HSO. Schriever also believed it served a useful purpose in precisely this way and was not simply "for show." Schriever, telephone interview with author, 4 March 1999, Grand Forks, N.D., tape recording at USAF HQ HSO.

21. John Lonnquest, "The Face of Atlas: General Bernard Schriever and the Development of the Atlas Intercontinental Ballistic Missile 1953–1960" (Ph.D. diss., Duke University, 1996), 258–59, 266–67, 291–97. Schriever, interview with author, 4 March

1999. Schriever believes that the idea for Black Saturday was his own, because his philosophy of management was to dig out the problems and not spend time on things that were going well. He did not recall seeing this practice on any other programs at that time. Schriever's group soon began to use computers for the management control system, with R-W. This would make it one of the earliest applications of computers to management communication.

22. Ethel M. DeHaven, *Air Materiel Command Participation in the Air Force Missiles Program through December 1957* (Inglewood, Calif.: Office of Information Services, Air Materiel Command Ballistic Missile Office, 1958), 185–87.

23. Letter to Major Franzel (RDGB) from General Schriever, Subject: Test philosophy for guided missiles, requesting General Powers' signature, 7 February 1955, AFHRA Microfilm 35257. This includes a memorandum to the Chief of Staff, HQ, USAF, from WDD HQ, ARDC, Subject: Test philosophy for guided missiles.

24. Ibid. Tab E, RW-EO(I)-011-56-e, January 1957, "The Role of Ramo-Wooldridge in the Ballistic Missile Testing Program," AFHRA Microfilm 35267.

25. *Symposium on Guided Missile Reliability, 2, 3, 4 November 1955,* sponsored by the Department of Defense, under auspices of Air Research and Development Command, AF-WP-O-21, September 1956, 56RDZ-12531, Wright-Patterson Air Force Base, Ohio, AFHRA Microfilm 26254.

26. Major Ernst Luke, Ernst Lange, Holloman Air Development Center, "Weapon System Evaluation and Reliability," in *Symposium on Guided Missile Reliability,* 169–70.

27. Jacob Neufeld, *Ballistic Missiles in the United States Air Force 1945–1960* (Washington, D.C.: Office of Air Force History, United States Air Force, 1990), 178–79, 205–6. Atlas Program Information Prepared for Stennis Sub-Committee, 20 May 1961, AFHRA K243.012-42, section on "Reliability Program."

28. Neufeld, *Ballistic Missiles,* 217–20. "Excerpts from Task Force's Report to Kennedy on U.S. Position in Space Race," *New York Times,* 12 January 1961. For details about Atlas failures, their causes, and responses to them, see "Problems of the Atlas Program: Status-Corrective Action," 26 April 1965, AFHRA Box J6 and J7, no file no. or name.

29. Letter to WDG from Colonel Soper, Subject: (U) missile reliability, accuracy, 1 April 1960, AFHRA Box J3–J5, file ICBM-1-19, Ops Reliability [1959].

30. Letter to General Ritland from General Terhune, Subject: AFFTC resources application to rocket development, 19 April 1960, AFHRA Box J3–J5, file ICBM-1-19, Ops Reliability [1959].

31. Neufeld, *Ballistic Missiles,* 217–19.

32. Robert L. Perry, *System Development Strategies: A Comparative Study of Doctrine, Technology, and Organization in the USAF Ballistic and Cruise Missile Programs, 1950–1960* RAND Corporation Memorandum RM-4853-PR (Santa Monica, Calif.: RAND Corporation, August 1966), 88.

33. Ethel M. DeHaven, *Air Materiel Command Participation,* 302–4.

34. Atlas Program Information for Stennis Subcommittee, "Reliability Program."

35. Ibid.

36. Terhune, interview with author, 24 September 1998. Simon Ramo, telephone interview with author, 14 February 2000, Grand Forks, N.D., notes with author. Bernard Schriever, telephone interview with author, 25 March and 13 April 1999, Grand Forks, N.D., tape recordings at USAF HQ HSO. J. D. Hunley, "The Evolution of Large Solid Propellant Rocketry in the United States," *Quest* 6, no. 1 (1998): 22–39.

37. Bellis, "Requirements for Configuration Management During Concurrency."

38. Lysle A. Wood, Boeing Company, "Configuration Management," in *AFSC Management Conference,* Monterey, Calif., 2–4 May 1962 (Washington, D.C.: Air Force Systems Command, 1962), 5-23-1–5-23-4.

39. Statement of Clyde Skeen, Assistant General Manager, Boeing Airplane Company, 30 April 1959, in *Weapons System Management and Team System Concept in Government Contracting,* Hearings before the Subcommittee for Special Investigations of the Committee on Armed Services, House of Representatives, 86th Congress, 1st Session (Washington, D.C.: Government Printing Office, 1959), 201–7.

40. Atlas Program Information for Stennis Subcommittee, "Reliability Program."

41. The Aerospace Corporation was a nonprofit corporation established to perform the same functions as STL. Its formation will be discussed later.

42. *System Requirements Analysis Orientation Guide* (San Bernardino, Calif.: Space Technology Laboratories, Thompson-Ramo-Wooldridge, ca. 1960), for the Minuteman Program, Office of U.S. Air Force Systems Command, LC/SPP-18.

43. Hall, interview with Neufeld, 11 July 1989. Hall was an abrasive character, a point clear from his autobiography and interviews with him. Importantly, his supervisors Schriever and Terhune believed this too and replaced him with the tactful Phillips. See author's interviews with Schriever and Terhune. See also, Edward N. Hall, *The Art of Destructive Management: What Hath Man Wrought?* (New York: Vantage Press, 1994).

44. "Finding Aid," n.d., LC/SPP-2.

45. The evidence points to the Minuteman project as the origin of the idea of tying financial controls to change control but does not definitely attribute that idea to Boeing or to the air force. What is clear is that Phillips quickly saw the importance of the idea and promoted it vigorously.

46. U.S. Air Force, Headquarters, Air Force Systems Command, *Air Force Systems Command Manual, Systems Management, AFSCM 375-1: Configuration Management during the Acquisition Phase,* 1 June 1962, LC/SPP-18.

47. See the U.S. Department of Transportation's launch vehicle Web site.

48. Air Force Regulation 70-9, "Procurement and Contracting," 12 November 1953, Marvin C. Demler Papers, AFHRA 168.7265-237. Air Force Ballistic Missile Management, *Formation of Aerospace Corporation,* Third Report by the Committee on Government Operations, 87th Congress, 1st Session, House Report No. 324 (Washington, D.C.: Government Printing Office, 1961), 4–7.

49. Neufeld, *Ballistic Missiles,* 111.

50. DeHaven, *Air Materiel Command Participation,* 184–86. *Formation of Aerospace Corporation,* 6–7.

51. Dyer, *TRW,* 198–206.

52. Ibid., 218–21.

53. MITRE Corporation was a nonprofit corporation in Bedford, Mass., created in 1958 to perform systems engineering for the Semi-Automatic Ground Equipment (SAGE) air-defense program.

54. Dyer, *TRW,* 228–32. *Formation of Aerospace Corporation,* 12–18.

55. Michael H. Gorn, *Vulcan's Forge: The Making of an Air Force Command for Weapons Acquisition (1950–1985)* (Washington, D.C.: Office of History, Air Force Systems Command, 1989), 63–69. Dennis J. Stanley and John J. Weaver, *An Air Force Command for R&D, 1949–1976: The History of ARDC/AFSC* (Washington, D.C.: Office of History, Headquarters, Air Force Systems Command, 1976), 34–37. The most detailed description of the entire Anderson Group effort is in Harry C. Jordan, *History of Ballistic Systems Division, Air Force Systems Command, The Division Pronaos,* Historical Division, Office of Information, Ballistic Systems Division, AF Systems Command, Norton AF Base, Calif., 3 April 1963, MITRE Box 1753.

56. Schriever, interview with author, 13 April 1999. Gorn, *Vulcan's Forge,* 69. Stanley and Weaver, *Air Force Command for R&D,* 37. Jordan, *History of Ballistic Systems Division,* 97.

57. Department of the Air Force, *Air Force Regulation No. 375-1, Systems Management, Management of Systems Programs,* 12 February 1962, LC/SPP-37, 1. Department of the Air Force, *Air Force Regulation No. 375-2, Systems Management, System Program Office,* 12 February 1962, LC/SPP-37. Department of the Air Force, *Air Force Regulation No. 375-3, Systems Management, System Program Director,* 12 February 1962, LC/SPP-37. Department of the Air Force, *Air Force Regulation No. 375-4, Systems Management, System Program Documentation,* 12 February 1962, LC/SPP-37. Department of the Air Force, *Air Force Regulation No. 80-5, Research and Development, Reliability Program for Systems, Subsystems, and Equipment,* 4 June 1962, LC/SPP-37.

58. See Gorn, *Vulcan's Forge,* 71–72. Stanley and Weaver, *Air Force Command for R&D,* 39–41. David N. Spires, *Beyond Horizons: A Half Century of Air Force Space Leadership* (Washington, D.C.: USAF, 1997), 86–92. Bernard A. Schriever, telephone interview with author, 16 June 1999, Grand Forks, N.D., tape recording at USAF HQ HSO.

59. Gorn, *Vulcan's Forge,* 72. Stanley and Weaver, *Air Force Command for R&D,* 41, 44.

60. U.S. Air Force, Air Force Systems Command, *Rainbow Reporting: Systems Data Presentation and Reporting Procedures, Program Management Instruction 1-5,* 15 January 1963, LC/SPP-18.

61. See Jacob Neufeld, ed., *Reflections on Research and Development in the United States Air Force: An Interview with General Bernard A. Schriever and Generals Samuel C.*

Phillips, Robert T. Marsh, and James H. Doolittle, and Dr. Ivan A. Getting, conducted by Dr. Richard Kohn (Washington, D.C.: Center for Air Force History, 1993), 65.

62. In his book *Inside Bureaucracy* (Boston: Little, Brown, 1967), Anthony Downs notes that organizations become more conservative over time and are brought under more control. The case supports Downs's theory. C. W. Borklund, *The Department of Defense* (New York: Frederick A. Praeger, 1968). John J. McLaughlin, "The Air Force's Management Revolution," *Air Force Magazine,* September 1962, 98–106.

63. Deborah Shapley, *Promise and Power, The Life and Times of Robert McNamara* (Boston: Little, Brown, 1993).

64. C. W. Borklund, *The Department of Defense* (New York: Frederick A. Praeger, 1968), 70–71. Charles J. Hitch, *Decision-Making for Defense* (Berkeley: University of California Press, 1967), 31–32. Memorandum to Research Council from F. T. Moore, Subject: The 92 Labors of Secretary McNamara, 17 March 1961, WM-663, CBI 90, Burroughs Collection, System Development Corporation Series, Box 1, RAND Corporation: Correspondence and Clippings, folder. McNamara initially instituted 92 projects, but these grew to more than 100—hence the number differences.

65. Shapley, *Promise and Power,* 99–101. James M. Roherty, *Decisions of Robert S. McNamara* (Coral Gables, Fla.: University of Miami Press, 1970), 72–73. Quotation from "DOD System Acquisition Policies," internal summary used for an air force meeting ca. 1962, LC/SPP-19.

66. Memorandum from John H. Rubel, Deputy Director of Defense Research and Engineering, to the Secretary of Defense, Subject: Management of research and engineering, 14 November 1961, LC/SPP-19.

67. Memorandum from John H. Rubel, Deputy Director of Defense Research and Engineering, to the Assistant Secretary of the Air Force (R&D), Subject: Standardized AGENA, 4 October 1961, LC/SPP-19. Memorandum from John H. Rubel, Deputy Director of Defense R&E, to the Assistant Secretary of the Air Force (R&D), Subject: Titan III Launch Vehicle Family, 13 October 1961, LC/SPP-19.

68. Memorandum from John H. Rubel, Deputy Director of Defense Research and Engineering, to the Assistant Secretary of the Air Force (Material) and the Assistant Secretary of the Air Force (R&D), Subject: Planning for Titan III Phase I efforts, 6 December 1961, LC/SPP-19.

69. Memorandum from John H. Rubel, for Harold Brown, Director of Defense Research and Engineering, to the Secretary of the Navy and the Secretary of the Air Force, Subject: Mobile mid-range ballistic missile program plan, 19 February 1962, LC/SPP-19.

70. See Merton J. Peck and Frederic M. Scherer, *The Weapons Acquisition Process: An Economic Analysis* (Cambridge: Division of Research, Graduate School of Business Administration, Harvard University, 1962), 19–24. Ants Kutzer, "Implementation of a Satellite Project," paper presented on the occasion of leaving the European Space Technology Centre on 15 September 1967, HAEUI Fond 51048, Table I.

71. Memorandum from Gen. B. A. Schriever, Commander AFSC, to Gen. C. E.

LeMay, HQ USAF, Subject: Systems acquisition management improvement, 5 February 1962, LC/SPP-19, 1.

72. Ibid., 2. Special Projects Office, "Program Evaluation Research Task, Summary Report, Phase 1," Bureau of Naval Weapons, Department of the Navy, Washington, D.C., July 1958. It was later renamed the Program Evaluation and Review Technique.

73. U.S. Air Force, Air Force Systems Command, *A Summary of Lessons Learned from Air Force Management Surveys,* AFSCP 375-2, 1 June 1963, LC/SPP-36. All quotations from Memorandum for distribution from Maj. General Robert J. Friedman, DCS/Comptroller, Subject: Guidance and plan for follow-through on the work of the Monterey Management Conference, 12 June 1962, LC/SPP-19. U.S. Air Force, Air Force Systems Command, *AFSC Management Objectives, FY 1964,* August 1963, LC/SPP-18.

74. See Viewgraphs for Air Force Institute of Technology System Program Management Course, probably presented by Samuel C. Phillips, ca. 1963, LC/SPP-37. U.S. Air Force, Air Force Systems Command, *Systems Management Newsletter,* no. 6, May 1963, LC/SPP-18. Memorandum to MSFC, KSC, MSC, from Maj. General Samuel C. Phillips, Deputy Director, Apollo Program, Subject: System program management course at Air Force Institute of Technology, 10 October 1964, LC/SPP-37.

75. Department of Defense Directive 3200.9, "Initiation of Engineering and Operational System Development," 1 July 1965.

76. William B. Bergen, "New Management Approach at Martin," *Aviation Age* 20, no. 6 (June 1954): 39–47; quotations at 39–40.

77. William B. Harwood, *Raise Heaven and Earth: The Story of Martin Marietta People and Their Pioneering Achievements* (New York: Simon and Schuster, 1993), 253, 278. Bergen later became manager of North American's Apollo program. Edward G. Uhl, "Applying the Systems Method to Air Weapons Development," *Aviation Age* 20, no. 2 (February 1954): 20–23.

78. Glenn E. Bugos, "Manufacturing Certainty: Testing and Program Management for the F-4 Phantom II," *Social Studies of Science* 23 (1993): 271–75.

79. Keith Davis, "The Role of Project Management in Scientific Manufacturing," *IRE Transactions on Engineering Management,* vol. EM-9, no. 3 (1962): 109–10 (emphasis in original).

80. H. F. Lanier, "Organizing for Large Engineering Projects," *Machine Design* 28 (1956): 54.

81. Ibid., 55–56.

82. Ibid., 57.

83. Ibid.

84. Charles S. Ames, "The Atlas Program at General Dynamics/Astronautics," in Fremont Kast and James Rosenzweig, eds., *Science, Technology, and Management* (New York: McGraw-Hill, 1963), 199–203; quotations at 203.

85. Ibid., 204.

86. This is an example of Alfred Chandler's idea that corporate structure follows corporate strategy. Alfred Chandler, *Strategy and Structure: Chapters in the History of Industrial Enterprise* (Cambridge: MIT Press, 1962).

Four | JPL's Journey from Missiles to Space

Appendices to Final Report of the Ranger Board of Inquiry, NASA, Washington, D.C., 30 November 1962, JPLA 2-2464, 15.

1. A version of this chapter has appeared as Stephen B. Johnson, "Craft or System? The Development of Systems Engineering at JPL," *Quest* 6, no. 2 (1998): 17–31.

2. Clayton Koppes, *JPL and the American Space Program: A History of the Jet Propulsion Laboratory* (New Haven: Yale University Press, 1982), 1–8. Frank J. Malina, "Origins and First Decade of the Jet Propulsion Laboratory," in Eugene M. Emme, ed., *The History of Rocket Technology* (Detroit: Wayne State University Press, 1964), 46–52.

3. Koppes, *JPL,* 9–11. Malina, "Origins and First Decade," 52–54. J. D. Hunley, "A Question of Antecedents: Peenemünde, JPL, and the Launching of U.S. Rocketry," in Roger D. Launius, ed., *Organizing for the Use of Space: Historical Perspectives on a Persistent Issue,* AAS History Series, vol. 18, R. Cargill Hall, series ed. (San Diego: Univelt for the AAS, 1995), 15–24.

4. Malina, "Origins and First Decade," 56–59. Koppes, *JPL,* 11–17. For the history of solid-propellant rocketry, see J. D. Hunley, "The Evolution of Large Solid Propellant Rocketry in the United States," *Quest* 6, no. 1 (1998): 22–38.

5. Koppes, *JPL,* 18–21, 26–30.

6. Malina, "Origins and First Decade," 60–66. Koppes, *JPL,* 22–24. WAC stood for Women's Auxiliary Corps (the "little sister of Corporal") or Without Attitude Control.

7. Hunley, "Evolution," 26. Tsien returned to China in 1955 and helped found the Chinese missile program. Iris Chang, *Thread of the Silkworm* (New York: Basic Books, 1995). Brian Harvey, *The Chinese Space Programme: From Conception to Future Capabilities* (New York: John Wiley & Sons, 1998).

8. F. J. Malina, "Development and Flight Performance of a High Altitude Sounding Rocket the 'WAC Corporal'," JPL Report No. 4-18, 24 January 1946, JPLA Historian Index No. 03 00066 XF.

9. Frank G. Denison Jr. et al., *Design and Development of the Corporal E,* Progress Report No. 4-61, 1 April 1949, ORDCIT Project Contract No. W-04-200-ORD-455, JPLA document D033, 20–35.

10. Denison et al., *Design and Development,* 36–37. The redesign used some ideas gleaned from von Braun's team, which was nearby at Fort Bliss. See Koppes, *JPL,* 39–40.

11. Dunn replaced Malina in 1947 when Malina resigned to work for the United Nations Educational, Scientific, and Cultural Organization.

12. Koppes, *JPL,* 40–44. See also Edmund Beard, *Developing the ICBM: A Study in Bureaucratic Politics* (New York: Columbia University Press, 1976).

13. Koppes, *JPL,* 31–32, 65.

14. Koppes, *JPL,* 46–52. "Freezing" a design meant not allowing changes without the permission of the project manager or the project manager's designate.

15. *Bimonthly Summary Report No. 23a, Supplement to Combined Bimonthly Summary No. 23, February 20, 1951, to April 20, 1951,* 15 May 1951, JPLA, 13. *Bimonthly Summary Report No. 37a, Supplement to Combined Bimonthly Summary No. 37, the Corporal Guided Missile XSSM-A-17,* 1 October 1953, JPLA, 38. *Bimonthly Summary Report No. 36a, Supplement to Combined Bimonthly Summary No. 36, the Corporal Guided Missile XSSM-A-17,* 1 August 1953, JPLA, 30.

16. *Bimonthly Summary Report No. 34a, Supplement to Combined Bimonthly Summary No. 34, the Corporal Guided Missile XSSM-A-17,* 1 April 1953, JPLA, 28–34; quotation at 29. Jack James, telephone interview with author, Grand Forks, N.D., 11 March 1997, notes with author.

17. *Bimonthly Summary Report No. 37a,* 36.

18. The IBM computer was most likely JPL's IBM 605. This was the only digital computer at JPL until July 1954. See Russell E. Carr et al., "The Digital-Computer Facility of the Jet Propulsion Laboratory," Report No. 20-99, ORDCIT Project, Contract No. DA-04-495-Ord 18, Dept. of the Army, Ordnance Corps, 1 October 1954, Item C025, Bibliography 39-1, *Publications of the JPL from January 1938 through June 1961,* JPLA, 1. *Bimonthly Summary Report No. 37a,* 36.

19. *Bimonthly Summary Report No. 37a,* 37–40.

20. Clarence R. Gates, Memorandum No. 20-76, "The Reliability of Redundant Systems," 27 August 1952, ORDCIT Project, Contract No. DA-07-495-ORD 18, Dept. of the Army, Ordnance Corps, JPLA document G009.

21. Clifford I. Cummings, "Some Army Problems in Missile Maintenance," External Publication No. 428, ORDCIT Project, Contract No. DA-04-495-Ord 18, Dept. of the Army, Ordnance Corps, 29 October 1957, Item C130, Bibliography 39-1, *Publications of the JPL from January 1938 through June 1961,* JPLA, 2–4. Koppes, *JPL,* 53–54.

22. J. N. James, *Planning the Instrumentation to Support an Integrated Weapon-System Development,* Publication No. 102, 6 June 1957, JPLA document J013, 2–3.

23. Koppes, *JPL,* 50–51, 56–57.

24. Ibid., 57–61.

25. Ibid., 62–65; quotation at 64. Dunn was a member of the Atlas "Teapot Committee" before joining R-W and hence helped accelerate and create the new institutional structure of the air force's Western Development Division. See also *JPL Sergeant Final Report,* 1960, for the Dept. of the Army, Ordnance Corps, Contract No. DA-04-495-Ord 18, JPLA 3-492-d, 4–5. Jack James stated that the Department of Defense came to JPL to run the ICBM program but that Caltech management refused the job, after which time Schriever selected Ramo-Wooldridge. Pickering's deputy, Colonel Terhune, also went to work for Schriever; Dunn went to Ramo-Wooldridge to run *Atlas,* as did JPL's radar expert Frank Lehan. James, interview with author, 11 March 1997.

26. *JPL Sergeant Final Report,* 142.

27. James, interview with author, 11 March 1997.

28. James, *Planning the Instrumentation,* 4–10, 15. J. N. James, *Development by the*

Jet Propulsion Laboratory of a Testing, Maintenance, and Supply Philosophy for the Sergeant System, Publication No. 104 (Revised), 8 August 1957, revised 14 February 1958, JPLA document J014.

29. James, *Planning the Instrumentation,* 4–10, 15. James, interview with author, 11 March 1997. Cummings, "Some Army Problems in Missile Maintenance," 6–11. J. N. James, *Development by the Jet Propulsion Laboratory.*

30. *Final Summary Report and Proposal for Technical Guidance Direction No. 511B Sergeant Weapon System,* Prepared for JPL, Contract No. DA-30-069-ORD-1783 by Sperry Utah Engineering Laboratory, SUEL Report No. 230-14070, April 1958, JPLA, 6-5–6-6. *JPL Sergeant Final Report,* 11–14, 143.

31. *JPL Sergeant Final Report,* 198.

32. James, interview by author, 11 March 1997. *JPL Sergeant Final Report,* 6–7.

33. *JPL Sergeant Final Report,* 6–7, 14–16, 203–5.

34. Koppes, *JPL,* 66–67, 80–82. *Jupiter Bimonthly Summary No. 1,* 1 August 1956, JPLA, 46–47. *Jupiter Bimonthly Summary No. 4, for the period 15 November 1956 to 15 January 1957,* 1 February 1957, 44–47, JPLA. *Jupiter Bimonthly Summary No. 8, for the period 15 July 1957 to 15 September 1957,* 1 October 1957, JPLA.

35. Constance McLaughlin Green, *Vanguard—A History,* SP-4202 (Washington, D.C.: NASA, 1970). William B. Harwood, *Raise Heaven and Earth: The Story of Martin Marietta People and Their Pioneering Accomplishments* (New York: Simon and Schuster, 1993). John Hagen, "The Viking and the Vanguard," in Emme, ed., *History of Rocket Technology,* 122–41.

36. Koppes, *JPL,* 85–86. *Jupiter Bimonthly Summary No. 10,* 1 February 1958, JPLA, 40–44. Pickering thought that Medaris might have von Braun's group develop the spacecraft, leaving JPL out in the cold.

37. *Jupiter Bimonthly Summary No. 1,* 83–84. G. Robillard, Publication No. 145, *The Explorer Rocket Research Program,* 31 October 1958, JPLA document R093, 4–5, 11–20. *Jupiter Bimonthly Summary No. 12,* 1 June 1958, JPLA, 54–58. *Space Research Summary No. 1, for the period 15 May 1958 to 15 July 1958,* 1 August 1958, JPLA, 9.

38. Robillard, *Explorer Rocket,* 12–14; *Space Research Summary No. 3, for the period 15 September 1958 to 15 November 1958,* 1 December 1958, JPLA, 6. *Space Research Summary No. 2, for the period 15 July 1958 to 15 September 1958,* 1 October 1958, JPLA, 20. *History of Ordnance Research at the Jet Propulsion Laboratory 1 January 1957 through 31 December 1958,* Publication No. 148, 6 February 1959, JPLA document J052, 14, 17–18. Henry Curtis and Dan Schneiderman, *Pioneer III and IV Space Probes,* 29 January 1960, Technical Release No. 34-11, JPLA document S065, 11.

39. Koppes, *JPL,* 98–103.

40. *Space Programs Summary No. 6, for the period 15 September 1959 to 15 November 1959,* 1 December 1959, NASA Contract No. NAS w-6, JPLA, 49–55. R. Cargill Hall, *Lunar Impact: A History of Project Ranger,* SP-4210 (Washington, D.C.: NASA, 1977), 15–17. Koppes, *JPL,* 103–4.

41. Cummings, "Some Army Problems in Missile Maintenance." Clifford I. Cum-

mings, Trip Report No. RPD-7, "Conference at Washington, D.C.," 5 April 1960, JPLA 2-1031. Memorandum from C. I. Cummings to Dr. W. H. Pickering, Subject: NASA Reliability Assessment Program, 4 November 1960, JPLA 2-1085.

42. *Space Programs Summary No. 6,* 11–12.

43. Koppes, *JPL,* 105–6. Hall, *Lunar Impact,* 17–24.

44. Hall, *Lunar Impact,* 23.

45. Ibid., 28–29, 34–35. *Improving Organizational Structure and Administrative Processes, Jet Propulsion Laboratory* (Los Angeles: McKinsey & Company, September 1959), chap. 1. James, interview with author, 11 March 1997. James described a minor "revolt" against Pickering's organization. Pickering agreed they could reorganize, but only if the managers worked on the reorganization on their own time. Cummings, "Some Army Problems in Missile Maintenance."

46. Hall, *Lunar Impact,* 28–29, 65–66.

47. *Appendices to Final Report of the Ranger Board of Inquiry,* NASA, Washington, D.C., 30 November 1962, JPLA 2-2464, 14–17.

48. Ibid., 16–18. Hall, *Lunar Impact,* chap. 3.

49. Hall, *Lunar Impact,* 25–30.

50. Ibid., 32–33, 39–43. *Space Programs Summary No. 37-7 for the period November 15, 1960 to January 15, 1961,* 1 February 1961, JPLA, 4–8.

51. Abe Silverstein, Director of Space Flight Programs, letter to Dr. William Pickering, date unreadable, ca. fall 1961, JPLA 3-361. *Space Programs Summary No. 37-12, Volume I, for the period September 1, 1961 to November 1, 1961,* 1 December 1961, JPLA, 18. Koppes, *JPL,* 127–28.

52. James, interview with author, 11 March 1997. *Mariner-Venus 1962 Final Project Report,* SP-59 (Washington, D.C.: NASA, 1965), JPLA 8-92, 21–24.

53. *Space Programs Summary No. 37-7,* 51–52. *Mariner-Venus 1962 Final Project Report,* 21–24. *From Project Inception through Midcourse Maneuver,* vol. 1 of *Mariner Mars 1964 Project Report: Mission and Spacecraft Development,* Technical Report No. 32-740, 1 March 1965, JPLA 8-28, 35. Memorandum from B. T. Morris to H. M. Schurmeier, Subject: Progress in failure reporting systems, 10 May 1962, JPLA.

54. *Space Programs Summary No. 37-11, Volume I, for the period July 1, 1961 to September 1, 1961,* 1 October 1961, JPLA, 3–4. *Space Programs Summary No. 37-13, Volume I, for the period November 1, 1961 to January 1, 1962,* 1 February 1962, 3–4. Hall, *Lunar Impact,* 99–111.

55. Hall, *Lunar Impact,* 138–47.

56. Ibid., 150–55, 164.

57. Koppes, *JPL,* 128. Hall, *Lunar Impact,* 160.

58. Hall, *Lunar Impact,* 164–66. *Space Programs Summary No. 37-17, Volume I, for the period July 1, 1962 to September 1, 1962, The Lunar Program,* NASA Contract No. NAS 7-100, 30 September 1962, JPLA.

59. *Space Programs Summary No. 37-21, Volume I, for the period March 1, 1963 to April 30, 1963, The Lunar Program,* 31 May 1963, JPLA, 2–3. Hall, *Lunar Impact,* 166–70. Koppes, *JPL,* 129.

60. *Appendices to Final Report of the Ranger Board of Inquiry,* 9–16, 146–47. NASA, *Final Report of the Ranger Board of Inquiry,* 30 November 1962, JPLA 2-2463, 5.

61. *Appendices to Final Report of the Ranger Board of Inquiry,* 15.

62. Ibid., 15, 19–20.

63. It is important to note that between 1958 and 1962, failures were considered part of the learning process. By 1962, NASA and Congress were becoming less tolerant of failure.

64. JPL, *Kelley Board Recommendations and Resultant Actions,* n.d., JPLA 2-2470.

65. Ibid. Hall, *Lunar Impact,* 176–80. R. C. Hall, *Systems Engineering, JPL Management Concepts, and the Ranger Project, A Review,* 21 February 1968, JPLA Historian Index No. 03 00791 XF, 7–8.

66. JPL, *Kelley Board Recommendations and Resultant Actions.* Memorandum to distribution from A. E. Wolfe, Subject: DEV testing, 7 November 1962, JPLA 2-1696. *Space Programs Summary No. 37-21, Volume I, for the period March 1, 1963 to April 30, 1963, The Lunar Program,* 31 May 1963, JPLA, 13–14. Hall, *Lunar Impact,* 186.

67. Memorandum to distribution from W. E. Giberson, Subject: December 4 Surveyor Project Review, 9 January 1964, JPLA 6-102. JPL, "Surveyor Project Review: March 23–27, 1964," JPLA 6-18b.

68. Hall, *Lunar Impact,* 235–38.

69. *Hilburn Board Recommendations and Resultant Actions,* date ca. March–April 1964, JPLA 2-2469a. Letter from W. H. Pickering to Homer E. Newell, Associate Administrator, Office of Space Science and Applications, 22 May 1964, JPLA 2-2469c. Hall, *Lunar Impact,* 240–49.

70. Koppes, *JPL,* 156–60. W. Henry Lambright, *Powering Apollo: James E. Webb of NASA* (Baltimore: Johns Hopkins University Press, 1995), 133–34. Hall, *Lunar Impact,* 252–54, 264–70, 351. *Investigation of Project Ranger,* U.S. Congress, Hearings before the Subcommittee on NASA Oversight of the Committee on Science and Astronautics, U.S. House of Representatives, 88th Congress, 2d Session, No. 3 (Washington, D.C.: Government Printing Office, 1964).

71. *Surveyor Block I Project Development Plan,* Rough Draft, 11 November 1963, JPLA 6-194. Memorandum from SL/Director, Lunar and Planetary Programs to S/Associate Administrator for Space Science and Applications, signed by Oran Nicks, Subject: JPL manpower requirements, JPLA, 2.

72. Quotations from JPL, "Surveyor Project Review: March 23–27, 1964," 55–57. *Project Surveyor: Report of the Subcommittee on NASA Oversight,* U.S. House of Representatives, 89th Congress, 1st Session (Washington, D.C.: Government Printing Office, 1965), 25.

73. Koppes, *JPL,* 176–77.

74. *Project Surveyor,* 23–25, 31–32.

75. Memorandum for the record, signed by Oran Nicks, Subject: Management discussion with Hughes Aircraft Company Officials, 12 June 1964, JPLA 6-23d. JPL Deputy Director's Office, *JPL Response to NASA Headquarters Surveyor Project Review of March 23–27, 1964,* 22 September 1964, Pasadena, Calif., JPLA 6-17b. Letter

from Dr. Newell to Dr. Pickering, 13 July 1964, JPLA 6-23e. Letter from Dr. Newell to Dr. Pickering, 14 July 1964, JPLA 6-23f. Letter from Dr. Pickering to Mr. Cortright, 14 July 1964, JPLA 6-23h. Letter from Mr. Cortright to Dr. Pickering, 4 August 1964, JPLA 6-23I. Quotation from Hall, *Systems Engineering*, 8.

76. *Project Surveyor*, 26–28.

77. *Project Surveyor*, 30–35.

78. Planetary spacecraft can be launched only when the positions of Earth and the target planet are in favorable locations—in the case of Mars, about every 18 months.

79. *From Project Inception*, 14, 25, 26.

80. Ibid., 28, 30, 31.

81. Ibid., 31–35.

82. System test is the group of final tests performed between the time that the spacecraft is first fully assembled and the time that the spacecraft is launched.

83. *From Project Inception*, 35–36.

84. Koppes, *JPL*, 169–72.

85. Ibid, 181–84. Hall, *Lunar Impact*, 289–306.

86. See the next chapter for the origins of work package management at North American.

87. JPL, *Surveyor Final Project Report*, part I, Project Description and Performance, Technical Report 32-1265 (Pasadena, Calif.: JPL, 1 July 1969), JPLA 6-256, 45–46, 57–58. Quotations from D. S. Liberman, F. A. Paul, and E. F. Grant, *Failure Reporting and Management Techniques in the Surveyor Program*, SP-6504 (Washington, D.C.: NASA, 1967), JPLA 6-67, 1–16.

88. *Mariner Venus 67, Project Policy and Requirements*, EPD-350, 18 May 1966, JPL, Caltech, JPLA 8-41, II-8, -9, III-1, IV-1–IV-9.

89. *Mariner Venus 67, Project Policy and Requirements*, III-1–III-5, III-17–III-20; quotation at III-3.

90. *Mariner Venus 67, Project Policy and Requirements*, Sections V, VI. Koppes, *JPL*, 196.

91. John Small, Deputy Chief, Systems Division, "Systems Engineering in Space Exploration," in *Seminar Proceedings, Systems Engineering in Space Exploration, May 1-June 5, 1963*, 1 June 1965, JPLA Historian Index No. 02 00221 XF, 1, 13–14.

92. W. Downhower, Chief Systems Design Section, "Systems Design"; C. R. Gates, Chief, Systems Division, "Systems Analysis"; Marshall Johnson, Chief, Space Flight Operations Section, "Space Flight Operations"; and John Small, Deputy Chief, Systems Division, "Program Engineering and Project Problems," in *Seminar Proceedings*, 15–32, 33–38, 39–44, 45–53.

Five | Organizing the Manned Space Program

T. Alexander, "The Unexpected Payoff of Project Apollo," *Fortune* 79, no. 7 (1969): 114.

1. The term "manned" space is anachronistic; women now frequently fly on space

missions. However, it was the term used at the time, and no women flew on American programs of the 1960s and 1970s.

2. Alex Roland, *Model Research, The National Advisory Committee for Aeronautics, 1915–1958,* SP-4103 (Washington, D.C.: NASA, 1985). Also see NASA histories of the NACA and NASA field centers.

3. Andrew J. Dunar and Stephen P. Waring, *Power to Explore: A History of Marshall Space Flight Center 1960–1990,* SP-4313 (Washington, D.C.: NASA, 1999).

4. Henry C. Dethloff, *Suddenly, Tomorrow Came: A History of the Johnson Space Center,* SP-4307 (Washington, D.C.: NASA, 1993), 35–51. Initially, the STG was to be moved to Goddard Space Flight Center, but its size and goals led to the creation of an entirely new center in Houston.

5. Interview of Robert R. Gilruth by Michael D. Keller, 26 June 1967, "Gilruth, Robert R. Bio" folder 000782, LEK 1/6/2, NASAHO.

6. Shirley Thomas, "Robert R. Gilruth," in *Men of Space: Profiles of the Leaders in Space Research, Development, and Exploration* (Philadelphia: Chilton Company, 1962), 38–74. "Gilruth, Robert R. Bio," 63.

7. Robert R. Gilruth, "From Wallops Island to Project Mercury, 1945–1958: A Memoir," in R. Cargill Hall, ed., *History of Rocketry and Astronautics,* AAS History Series, vol. 7, part II (San Diego: AAS Publications Office, 1986), 445–76. Announcement, Key NASA Personnel Change, Director, Key Personnel Development, Director Manned Spacecraft Center, 18 January 1972, "Gilruth, Robert R. Bio" folder 000782, LEK 1/6/2, NASAHO.

8. Arnold S. Levine, *Managing NASA in the Apollo Era,* SP-4102 (Washington, D.C.: NASA, 1982), 61–64. Loyd S. Swenson Jr., James M. Grimwood, and Charles C. Alexander, *This New Ocean: A History of Project Mercury,* SP-4201 (Washington, D.C.: NASA, 1966), 120–22, 137.

9. *Project Mercury Status Report No. 2 for Period Ending April 30, 1959,* NASA, Langley STG, NASAHO, 1. *Project Mercury Quarterly Status Report No. 6 for Period Ending April 30, 1960,* NASA, STG, NASAHO, 33–34.

10. John M. Logsdon, Moderator, *Managing the Moon Program: Lessons Learned from Project Apollo,* Proceedings of an Oral History Workshop Conducted July 21, 1989, Monograph in Aerospace History No. 14 (Washington, D.C.: NASA, July 1999), 18.

11. *Mercury Project Summary, Including Results of the Fourth Manned Orbital Flight,* SP-45 (Washington, D.C.: NASA, 1963), 18–21.

12. Logsdon, *Managing the Moon Program,* 32–33.

13. *Mercury Project Summary,* 18–21.

14. Marlowe Cassetti, interview by Carol Butler, *Quest* 7, no. 1 (1999): 46.

15. Swenson et al., *This New Ocean,* 149, and *Project Mercury Status Report.*

16. Swenson et al., *This New Ocean,* 146–51, 153, 251.

17. Swenson et al., *This New Ocean,* 77–82. Courtney G. Brooks, James M. Grimwood, and Loyd S. Swenson Jr., *Chariots for Apollo: A History of Manned Lunar Spacecraft,* SP-4205 (Washington, D.C.: NASA, 1979), 4, 7, 8, 11, 12.

18. Roger Bilstein, *Stages to Saturn: A Technological History of the Apollo/Saturn Launch Vehicle,* SP-4206 (Washington, D.C.: NASA, 1980), 26–31, 79–81.

19. Brooks et al., *Chariots for Apollo,* 14–17. Ivan D. Ertel and Mary Louise Morse, *The Apollo Spacecraft: A Chronology,* vol. 1 (through 7 November 1962), SP-4009 (Washington, D.C.: NASA, 1969), 47–48, 54, 57–60.

20. John M. Logsdon, *The Decision to Go to the Moon: Project Apollo and the National Interest* (Cambridge: MIT Press, 1970). Stephen J. Garber, "Multiple Means to an End: A Reexamination of President Kennedy's Decision to Go to the Moon," *Quest* 7, no. 2 (1999): 5–17.

21. Ertel and Morse, *Apollo Spacecraft,* vol. 1, 54, 101–4, 120–27. Brooks et al., *Chariots for Apollo,* 35, 38–41.

22. Bilstein, *Stages to Saturn,* 81–82, 140–41. The *S-I* was a test stage, not the same as the *S-IB,* developed for the *Saturn V* lunar vehicle by Boeing. North American Rocketdyne was North American's engine group that was also working on the *F-1* engine.

23. Bilstein, *Stages to Saturn,* 267. Dunar and Waring, *Power to Explore,* 47–48.

24. Dunar and Waring, *Power to Explore,* 153.

25. Bilstein, *Stages to Saturn,* 262–63. Phillip K. Tompkins, *Organizational Communication Imperatives: Lessons of the Space Program* (Los Angeles: Roxbury, 1992), chaps. 3–5. Dunar and Waring, *Power to Explore,* 47.

26. STG, *Project Development Plan for Rendezvous Development Utilizing the Mark II Two-Man Spacecraft,* 27 October 1961, in James M. Grimwood, Barton C. Hacker, and Peter J. Vorzimmer, *Project Gemini: A Chronology,* SP-4002 (Washington, D.C.: NASA, 1969), 14–15. Barton C. Hacker and James M. Grimwood, *On the Shoulders of Titans: A History of Project Gemini,* SP-4203 (Washington, D.C.: NASA, 1977), chaps. 2 and 3, 80–81.

27. Engineers originally planned for *Gemini* to land on solid ground using a "paraglider," similar to today's hang gliders.

28. Hacker and Grimwood, *On the Shoulders of Titans,* 82, 122.

29. Mary Louise Morse and Jean Kernahan Bays, *The Apollo Spacecraft, A Chronology,* vol. 2 (8 November 1962–30 September 1964), SP-4009 (Washington, D.C.: NASA, 1973), 70–71. Brooks et al., *Chariots for Apollo,* 121.

30. C. D. Stroud Briefing on Program Management Systems, "The Boeing Company Apollo/Saturn V Program Briefing, New Orleans, La., October 17, 1968," in *Apollo Program Management,* Staff Study for the Subcommittee on NASA Oversight of the Committee on Science and Astronautics, U.S. House of Representatives, 91st Congress, 1st Session, July 1969 (Washington, D.C.: Government Printing Office, 1969), 19–22.

31. Howard McCurdy, *Inside NASA* (Baltimore: Johns Hopkins University Press, 1993), 30–33.

32. C. D. Stroud, "S-IC Program Management Systems," in *Apollo Program Management,* 19–20. Bilstein, *Stages to Saturn,* 194–95. Lt. General Samuel Phillips, interview with Thomas W. Ray and Loyd S. Swenson Jr., 22 July 1970, JSC Archives Box 074-51, 27–28. Dunar and Waring, *Power to Explore,* 86–87.

33. Bilstein, *Stages to Saturn,* 277–78; quotation at 277. Brooks et al., *Chariots for Apollo,* 112–14.

34. Brooks et al., *Chariots for Apollo,* 119–20.

35. Lt. General Samuel Phillips, United States Air Force, interview with Tom Ray, 22 July 1970, NASA, "Phillips, Samuel C., "Interviews" folder 001701, LEK 1/13/4, NASAHO, 34–39. Morse and Bays, *Apollo Spacecraft,* vol. 2, 12, 76–77.

36. Brooks et al., *Chariots for Apollo,* 120–21.

37. See "Survey of Man-Rating Criteria," author unknown, n.d., ca. June–August 1964, "Notes Jun–Aug 1964," LC/SPP-43:9. Mitchell R. Sharpe and Bettye B. Burkhalter, "Mercury-Redstone: The First American Man-Rated Space Launch Vehicle," in John Becklake, ed., *History of Rocketry and Astronautics,* AAS History Series, vol. 17, R. Cargill Hall, series ed. (San Diego: AAS Publications Office, 1995), 341–88.

38. *Project Mercury Quarterly Status Report No. 6 for Period Ending April 30, 1960,* NASA, STG, NASAHO, 12. William B. Harwood, *Raise Heaven and Earth: The Story of Martin Marietta People and Their Pioneering Accomplishments* (New York: Simon and Schuster, 1993), 373–78. Swenson et al., *This New Ocean,* 254–56.

39. Owen G. Morris, JSC Oral History Project Interview with Summer Chick Bergen, 30 June 1999.

40. *Project Mercury Quarterly Status Report No. 6,* 12, 28. *Project Gemini Status Report No. 1 for Period Ending May 31, 1962,* NASA, MSC, NASAHO, 4, 5, 33–34. Harwood, *Raise Heaven and Earth,* 373–78. *Lockheed Agena Monthly Report, June 1964,* 3–11, in Grimwood et al., *Project Gemini,* 146. *Project Mercury Quarterly Status Report No. 7 for Period Ending July 31, 1960,* NASA, STG, NASAHO, 26.

41. Swenson et al., *This New Ocean,* 148–49, 270, and *Project Mercury Status Report No. 2 for Period Ending April 30, 1959,* NASA, Langley STG, NASAHO, 23–31. Brooks et al., *Chariots for Apollo,* 138. Morse and Bays, *Apollo Spacecraft,* vol. 2, 87. Harmon L. Brendle and James A. York, *Apollo Experience Report—The Command and Service Module Milestone Review Process,* NASA TN D-7599, February 1974, "Apollo Experience Reports" folder 007822, NASAHO, 1.

42. Bilstein, *Stages to Saturn,* chap. 4. John J. Williams and Donald M. Corcoran, "Mercury Spacecraft Pre-Launch Preparations at the Launch Site," paper presented at the American Institute of Aeronautics and Astronautics Space Flight Testing Conference, Cocoa Beach, Fla., 18–20 March 1963, 18. Swenson et al., *This New Ocean,* 180–82. McCurdy, *Inside NASA,* 32–33.

43. *Project Mercury Quarterly Status Report No. 6,* 5–11. *Project Mercury Quarterly Status Report No. 11 for Period Ending July 31, 1961,* NASA, MSC, NASAHO. *Project Mercury Quarterly Status Report No. 10 for Period Ending April 30, 1961,* NASA, STG, NASAHO.

44. Swenson et al., *This New Ocean,* 270–71.

45. *Project Mercury Quarterly Status Report No. 6,* 33–34. Grimwood et al., *Project Gemini,* 37, 68, 146. Morse and Bays, *Apollo Spacecraft,* vol. 2, 65.

46. Swenson et al., *This New Ocean,* 275–79, 308, 321–22.

47. Swenson et al., *This New Ocean,* 294–301. Dunar and Waring, *Power to Explore,* 81.

48. Memorandum from North to Director, Space Flight Programs, "Mercury Capsule Changes and Flight Schedule," in Swenson et al., *This New Ocean,* 299.

49. *Project Mercury Quarterly Status Report No. 15 for Period Ending July 31, 1962,* NASA, MSC, NASAHO, 26. *Project Mercury Quarterly Status Report No. 16 for Period Ending October 31, 1962,* NASA, MSC, NASAHO, 22.

50. Grimwood et al., *Project Gemini,* 94, 136, 164–65. *Gemini Program/Project Development Plan,* Gemini Program Office, M-D MG 600-1, 1 July 1966, NASA Office of Manned Space Flight, 7-4. Brooks et al., *Chariots for Apollo,* 123.

51. Jacob Neufeld, *Ballistic Missiles in the United States Air Force 1945–1960* (Washington, D.C.: Office of Air Force History, United States Air Force, 1990), 218.

52. Swenson et al., *This New Ocean,* 643. *Mercury Project Summary,* 4–5. Hacker and Grimwood, *On the Shoulders of Titans,* 96–115.

53. Hacker and Grimwood, *On the Shoulders of Titans,* 77.

54. W. Henry Lambright, *Powering Apollo: James E. Webb of NASA* (Baltimore: Johns Hopkins University Press, 1995), 114–17.

55. Hacker and Grimwood, *On the Shoulders of Titans,* 128–29, 581. *Project Gemini Status Report No. 7 for Period Ending November 30, 1963,* NASA, MSC, NASAHO, 84–85. Chamberlin's removal resulted from Seamans's embarrassment when he and McNamara visited MSC and found the program in far worse technical and budgetary shape than either were aware of. Cost increases were rampant in the *Titan II* launch vehicle, *Agena,* the paraglider, fuel cells, and the orbital maneuvering system. Originally cast as a minor modification of *Mercury, Gemini* ultimately used little of the *Mercury* design.

56. Lambright, *Powering Apollo,* 120–22. Levine, *Managing NASA,* 81.

57. *Preliminary Report to the Associate Administrator on Studies Relating to Management Effectiveness in Scheduling and Cost Estimating NASA Projects,* by Deputy Associate Administrator, 15 September 1964, NASAHO.

58. The DOD conversions to phased planning started based on the Bell Report of 1961. *Report to the President on Government Contracting for Research and Development,* Executive Office of the President, Bureau of the Budget, Washington, D.C., 30 April 1962 (Bell Report), in John M. Logsdon, ed., *Exploring the Unknown: Selected Documents in the History of the U.S. Civil Space Program,* SP-4218 (Washington, D.C.: NASA, 1996), 652–53. *Preliminary Report to the Associate Administrator.*

59. Levine, *Managing NASA,* 81, 158–60. *Apollo Program Weekly Status Report for Nov. 2–6, 1964,* from MAP-1/Chief, Program Planning, 6 November 1964, LC/SPP-41.

60. Biographical sketch of Mueller, in *Apollo Accident Hearings,* 8 parts, 90th Congress, 1st and 2d Sessions, 7 February 1967–January 1968, Senate Committee on Aeronautical and Space Sciences, Part 1, 68.

61. Interview with George E. Mueller, by Putnam, 27 June 1967, "George E. Mueller Interviews" folder 001522, LEK 1/12/1, NASAHO, 6–7.

62. Quotations from Mueller, interview by William P. Putnam, 12–13. Dunar and Waring, *Power to Explore,* 66.

63. Lambright, *Powering Apollo,* 117–18. Brooks et al., *Chariots for Apollo,* 129. Mueller, interview by Putnam, 8, 13–14.

64. Lambright, *Powering Apollo,* 117–18. Bilstein, *Stages to Saturn,* 348-50. Morse and Bays, *Apollo Spacecraft,* vol. 2, 104.

65. McCurdy, *Inside NASA,* 94–96. Bilstein, *Stages to Saturn,* 350–51. The "all-up" concept succeeded because von Braun's experienced team could anticipate most of the problems and plan tests accordingly.

66. "GEM" stood for George E. Mueller.

67. Logsdon, *Managing the Moon Program,* 19. Quotation from Owen Moriss.

68. *An Evaluation of Organization and Management, Office of Manned Space Flight, National Aeronautics and Space Administration,* prepared by Management Analysis Division, NASA headquarters, August 1964, "notes Jun–Aug 1964" LC/SPP-43:9, III-2.

69. Memorandum from George E. Mueller to NASA Administrator, 26 September 1963, Subject: Utilization of Air Force Program Management personnel, "Notes Jan-May" LC/SPP-43:8. Phillips, interview with Ray and Swenson, 41.

70. Jacob Neufeld, ed., *Reflections on Research and Development in the United States Air Force: An Interview with General Bernard A. Schriever and Generals Samuel C. Phillips, Robert T. Marsh, and James H. Doolittle, and Dr. Ivan A. Getting,* conducted by Dr. Richard H. Kohn (Washington, D.C.: Center for Air Force History, 1993), 87.

71. S. C. Phillips, Deputy Director, Apollo Program, letter to General B. A. Schriever, Commander AFSC, 20 January 1964, "NASA Personnel-Military" folder LC/SPP-43:15. W. E. Leonhard, Chief of Staff, AFSC, letter to Brig. Gen. S. C. Phillips, NASA, 14 February 1964, "NASA Personnel-Military" folder LC/SPP-43:15. Samuel C. Phillips, notes on fifty-five NASA positions to be filled by air force officers, ca. March 1964, "Notes, Jan-May" LC/SPP-43:8. Hugh L. Dryden, Deputy Administrator, NASA, letter to Eugene M. Zuckert, Secretary of the Air Force, 1 April 1964, "NASA Personnel-Military" folder LC/SPP-43:15. Eugene M. Zuckert, letter to Hugh L. Dryden, Deputy Administrator, NASA, 5 May 1964, "NASA Personnel-Military" folder LC/SPP-43:15. Hugh L. Dryden, letter to Eugene M. Zuckert, Secretary of the Air Force, 12 May 1964, "NASA Personnel-Military" folder LC/SPP-43:15. *Report of Joint Air Force/NASA Military Requirements Review Group,* September 1964, "NASA Personnel-Military" folder LC/SPP-43:15. Phillips, interview with Ray and Swenson, 15–16.

72. James E. Webb, letter to General J. McConnell, Vice Chief of Staff, USAF, 16 December 1964, "NASA 1964 Personnel-Military" LC/SPP-43:15. J. McConnell, General, USAF, Vice Chief of Staff, letter to Dr. Robert C. Seamans Jr., 21 December 1964, "NASA 1964 Personnel-Military" LC/SPP-43:15. Memorandum from David M. Jones (?), Brig. Gen., USAF, Deputy Associate Administrator for Manned Space Flight (Programs), to General McKee, Subject: Job descriptions for Colonels O'Connor, Bolender, and Yarchin, 15 January 1965, LC/SPP-43:15.

73. Memorandum for record from Samuel C. Phillips, 11 February 1964, "Chrono-

logical File 1964" LC/SPP 41:2. Quotation from Samuel C. Phillips, notes, 29 February 1964, "Organization and Management, Conference in Los Angeles" LC/SPP-43:12.

74. Phillips, notes, 29 February 1964. Quotation from Samuel C. Phillips, notes and speech, ca. April 1964, "Organization and Management" LC/SPP-43:13.

75. Interoffice correspondence from C. W. Besserer, STL, to Brig. Gen. S. C. Phillips, Subject: Minuteman Control Room, 24 January 1963, "AF BSD Minuteman Correspondence 1963" LC/SPP-10:8. "Miscellaneous Viewgraphs," ca. January–February 1964, "Organization and Management" LC/SPP-43:11.

76. Bilstein, *Stages to Saturn,* 284–86. Tompkins, *Organizational Communication Imperatives,* 78–79. Dunar and Waring, *Power to Explore,* 67–68.

77. *Apollo Configuration Management Manual,* NPC 500-1, Office of Manned Space Flight, Apollo Program, NASA, 18 May 1964, "Config Mgmt Manual" LC/SPP-41:3. *Apollo Executives' Meeting Proceedings,* Asheville, N.C., 18–19 June 1964, "Executives Meeting" LC/SPP-42:6, III-8. Memorandum from Samuel C. Phillips to Marshall Space Flight Center, Apollo Program Manager, Subject: Apollo Configuration Management Manual, 25 May 1964, "Config Mgmt Manual" LC/SPP-41:3. R. B. Young, Director, Industrial Operations, MSFC, Letter to Maj. General Samuel C. Phillips, Subject: Apollo Configuration Management Manual, 19 June 1964, "Config Mgmt Manual" LC/SPP-41:3.

78. Memorandum to distribution from E. L. Harkleroad, Subject: Configuration management—development of Apollo specifications, 20 July 1964, LC/SPP-43:9.

79. *Configuration Management Provisional Group Report, SID 64-1608,* 15 September 1964, North American Aviation, Space and Information Systems Division, LC/SPP-41, 1, 25, 59–63; quotation at 25.

80. Col. John V. Patterson Jr., letter to Brig. Gen. S. C. Phillips, Director, Minuteman SPO, AFSC/BSD, 7 April 1963, LC/SPP-37. Memorandum from Maj. General Samuel C. Phillips, Deputy Director, Apollo Program, to MSFC, KSC, MSC, Subject: System management course at Air Force Institute of Technology, 10 October 1964, LC/SPP-37. See also memorandum from Maj. General Samuel C. Phillips to Admiral Boone, Subject: Defense Weapon Systems School, 1 December 1964, LC/SPP-41. This memorandum shows that two people from MSFC and two from HQ went to the class.

81. *Apollo Executives' Meeting Proceedings,* III-11, 13.

82. Ibid., III-16, 17.

83. Ibid., III-18.

84. Ibid., III-19.

85. Ibid., III-20.

86. Memorandum to file by Stanley Winn and Lt. Col. C. Taylor, Subject: Trip Report to MSC, Houston, on October 6 and 7, 1964, "Notes Sept-Dec 1964" LC/SPP-43:10. Quotation from Samuel C. Phillips, "Configuration Control," speech at Southeastern Symposium on Government Contracts, Cocoa Beach, Fla., 19 March 1965, LC/SPP-55:8, 1. Lee B. James, Manager, Saturn I/IB Program, MSFC, 26 August 1965, "Letter to Maj. General Samuel C. Phillips" LC/SPP-55:3.

87. "Common Industry Problems Found among Apollo Contractors," n.d., ca. mid-1965, "Systems Management" LC/SPP-55:3.

88. *NASA-Apollo Program Management,* vol. 1 (Washington, D.C.: NASA OMSF Apollo Program Office, December 1967), NASAHO, 4-44–4-46. *Apollo Program Directive No. 6A, Subject: Sequence and Flow of Hardware Development and Key Inspection, Review and Certification Checkpoints,* 30 August 1966, in *NASA-Apollo Program Management,* vol. 1, December 1967, app. J.

89. Courtney G. Brooks and Ivan D. Ertel, *The Apollo Spacecraft, A Chronology,* vol. 3, 1 October 1964–20 January 1966, SP-4009 (Washington, D.C.: NASA, 1976), 51, 97.

90. *NASA-Apollo Program Management,* vol. 1, 4-28–4-34.

91. Mike Gray, *Angle of Attack: Harrison Storms and the Race to the Moon* (New York: Penguin, 1992), 196–98. Dunar and Waring, *Power to Explore,* 89–90.

92. Bilstein, *Stages to Saturn,* 224–25.

93. "Tiger team" was an Air Force term for an ad hoc group to dig up and fix critical problems using the best available personnel. Bilstein, *Stages to Saturn,* 225. Phillips, interview with Ray and Swenson, 29–30.

94. Phillips to Mueller, quoted in Lambright, *Powering Apollo,* 152. Gray, *Angle of Attack,* 200–202.

95. Gray, *Angle of Attack,* 207–9. Bilstein, *Stages to Saturn,* 227–29. "Grumman Aircraft Engineering Corporation, Summary of the Report for the Subcommittee on NASA Oversight, November 4, 1968," in *Apollo Program Management,* House Staff Study, 45. See also "North American Rockwell Corporation," in *Apollo Program Management,* 175, 188. Harvard Business School professor J. Sterling Livingston knew about work package management in 1962. However, it did not reach NASA until North American accelerated its deployment at this time. See J. Sterling Livingston, Harvard University, "Advanced Techniques for Program/Funds Management," in *AFSC Management Conference,* Monterey, Calif., 2–4 May 1962 (Washington, D.C.: Air Force Systems Command, 1962).

96. "North American Rockwell Corporation," 174, 189.

97. Brooks et al., *Chariots for Apollo,* 197–201. Ivan D. Ertel and Roland W. Newkirk, with Courtney G. Brooks, *The Apollo Spacecraft: A Chronology,* vol. 4, 21 January 1966–13 July 1974, SP-4009 (Washington, D.C.: NASA, 1978), 27.

98. *Report of Apollo 204 Review Board to the Administrator,* NASA, 1967, by Floyd L. Thompson, chairman, 5 April 1967, with Appendices A–G, NASAHO. Gray, *Angle of Attack,* 226–31.

99. See Lambright, *Powering Apollo,* chaps. 8 and 9, for an excellent description of Webb's role in the aftermath of the fire.

100. *Apollo Accident Hearings,* 233. Gray, *Angle of Attack,* 232–35.

101. Lambright, *Powering Apollo,* 151–58.

102. Memorandum from James E. Webb, Administrator NASA, for Mr. Shapley/ADA, Subject: Problem of adding "supervision" to the activities of top management

to see that the plans made are carried out, both in the area of substance and in the area of administration, 19 September 1967, NASAHO.

103. Lambright, *Powering Apollo,* 161. James E. Webb, Administrator NASA, memorandum to Mr. Finger, 27 October 1967, NASAHO.

104. Lambright, *Powering Apollo,* 170–75. Gray, *Angle of Attack,* 244–56.

105. Lambright, *Powering Apollo,* 175–76. Phillips, interview with Tom Ray, 39–41. Phillips knew Boeing well because it was the Minuteman contractor.

106. See "The Boeing Company Apollo/Saturn V Program Briefing," in *Apollo Program Management,* House Staff Study, 31–34.

107. "Single point failure" is a term meaning any single fault that could cause a loss of the mission or loss of life. In general, it is impossible to eliminate all single-point failures from a design (the spacecraft structure is a typical example). Therefore, NASA found it necessary to identify each one and ensure that the probability of its occurrence was very small.

108. Ertel and Newkirk, *Apollo Spacecraft,* vol. 4, 152–53, 163, 176–78, 198. Memorandum from Sam C. Phillips to Mueller, 5 September 1968, George E. Mueller Correspondence folder 001553, LEK 1/12/1, NASAHO. This list is not exhaustive.

109. "North American Rockwell Corporation," 150–54.

110. Ertel and Newkirk, *Apollo Spacecraft,* vol. 4, 203–5, 223–24. Letter from George E. Mueller, NASA OMSF, to Robert R. Gilruth, MSC, 14 February 1968, NASAHO. Memorandum from Mueller to Phillips, Subject: Software Task Force meetings, 19 February 1968, NASAHO.

111. Ertel and Newkirk, *Apollo Spacecraft,* vol. 4, 250, 257.

112. Dunar and Waring, *Power to Explore,* chap. 5.

113. Tompkins, *Organizational Communication Imperatives,* 105–11.

114. Ibid., 112.

Six | Organizing ELDO for Failure

R. Aubinière, "Tenth Anniversary of the Establishment of ELDO," *ESRO/ELDO Bulletin* 24 (March 1974): 13.

1. Alfred Grosser, *The Western Alliance: European American Relations since 1945,* trans. Michael Shaw (New York: Vintage Books, 1982), from *Les Occidentaux, Les pays d'Europe et les Etats-Unis depuis la guerre* (Paris: Editions Fayard, 1978). Alan S. Milward, *The Reconstruction of Western Europe 1945–51* (Berkeley: University of California Press, 1984), and *The European Rescue of the Nation-State* (Berkeley: University of California Press, 1992). Charles Maier, *In Search of Stability* (Cambridge: Cambridge University Press, 1987).

2. Armin Hermann et al., *History of CERN* (Amsterdam: North-Holland Physics Publishing, 1987), 2 vols. Jaroslav G. Polach, *EURATOM* (Dobbs Ferry, N.Y.: Oceana Publications, 1964).

3. Jean-Jacques Servan-Schreiber, *The American Challenge,* trans. Robert Steel (New York: Avon Books, 1969), 168 (emphasis in original).

4. Robert McNamara, *The Essence of Security* (New York: Harper & Row, 1968), 109, 112.

5. Franz Josef Strauss, *Challenge and Response: A Programme for Europe,* trans. Henry Fox (New York: Atheneum, 1970), 126–29, 136–37.

6. McNamara, *Essence of Security,* 109.

7. Servan-Schreiber, *American Challenge,* 151 (emphasis in original).

8. Antonie T. Knoppers, "The Cause of Atlantic Technological Disparities," in Richard H. Kaufman, ed., *The Technology Gap: U.S. and Europe,* prepared by the Atlantic Institution (New York: Praeger Publishers, 1970), 134.

9. C. Freeman and A. Young, *The Research and Development Effort: Western Europe, North America, and the Soviet Union* (Paris: OECD, 1965). Organization for Economic Cooperation and Development, *Government and Technical Innovation* (Paris: OECD, 1966). Daniel Lloyd Spencer, *Technology Gap in Perspective* (New York: Spartan Books, 1970), 12–13.

10. "'Technology Gap' Studied by NATO," *New York Times,* 2 March 1967, "U.S.-Europe 1965–1972" folder 014548, LEK 7/9/4, NASAHO. "Technology Gap Upsets Europe," *New York Times,* 12 March 1967, "U.S.-Europe 1965–1972" folder 014548, LEK 7/9/4, NASAHO. "Europe is Said to Accept Fact of a 'Technology Gap,'" *New York Times,* 21 July 1967, "U.S.-Europe 1965–1972" folder 014548, LEK 7/9/4, NASAHO. "West Europeans Attribute Continuing Technology Lag Behind the U.S. to Inferior Management," *New York Times,* 13 December 1967, "U.S.-Europe 1965–1972" folder 014548, LEK 7/9/4, NASAHO. Robert Gilpin, *France in the Age of the Scientific State* (Princeton, N.J.: Princeton University Press, 1968). Henry R. Nau, "A Political Interpretation of the Technology Gap Dispute," *Orbis* 15, no. 2 (1971): 507–27.

11. Michael J. Neufeld, *The Rocket and the Reich: Peenemünde and the Coming of the Ballistic Missile Era* (New York: Free Press, 1995). Benjamin King and Timothy Kutta, *Impact: The History of Germany's V-Weapons in World War II* (Rockville Centre, N.Y.: Sarpedon, 1998).

12. Harrie Massey and M. O. Robins, *History of British Space Science* (Cambridge: Cambridge University Press, 1986), 11–12. Jean Corbeau, "A History of the French Sounding Rocket Veronique," in Kristan R. Lattu, ed., *History of Rocketry and Astronautics,* AAS History Series, vol. 8, R. Cargill Hall, series ed. (San Diego: AAS, 1989).

13. Corbeau, "History," 147–67; Claude Carlier and Marcel Gilli, *The First Thirty Years at CNES, The French Space Agency, 1962–1992* (Paris: La Documentation Française/CNES, 1994), 6.

14. Guy Collins, *Europe in Space* (New York: St. Martin's Press, 1990), 13–14. Jean-Pierre Causse, "Les lanceurs européens avant Ariane," in Emmanuel Chadeau, ed., *L'ambition technologique: Naissance d'Ariane* (Paris: Editions Rive Droite, 1995), 15–17. Berry Sanders, "The French *Diamant* Rockets," *Quest* 7, no. 1 (1999): 18–22. Col. Edward N. Hall, Oral History Interview with Jack Neufeld, 11 July 1989, AFHRA K239.0512-1820, 1–2, 26–29. Secretary of the Air Force Donald Quarles sponsored Hall's work in Europe. I do not know if others in the U.S. government knew of Hall's mission.

15. See John Krige, *The Launch of ELDO,* HSR-7 (Noordwijk, The Netherlands: ESA, 1993), 2–4. Stephen Robert Twigge, *The Early Development of Guided Weapons in the United Kingdom, 1940–1960* (Chur, Switzerland: Harwood Academic Publishers, 1993), 23–24, 40–43, 188–91. John Krige and Arturo Russo, *Europe in Space 1960–1973,* SP-1172 (Noordwijk, The Netherlands: ESA, 1994), 29. J. Krige and A. Russo, with contributions by M. De Maria and L. Sebesta, *A History of the European Space Agency 1958–1987,* vol. 1, SP-1235 (Noordwijk, The Netherlands: ESA, 2000).

16. Krige, *Launch of ELDO,* 4–6. Twigge, *Guided Weapons,* 193–202.

17. Krige, *Launch of ELDO,* 6–10.

18. Krige, *Launch of ELDO,* 11–17. Walter A. McDougall, "Space-Age Europe: Gaullism, Euro-Gaullism, and the American Dilemma," *Technology and Culture* 26 (1985): 179–203. Gilpin, *France in the Age.*

19. Krige, *Launch of ELDO,* 11–17. Michelangelo De Maria, *Europe in Space: Edoardo Amaldi and the Inception of ESRO,* HSR-5 (Noordwijk, The Netherlands: ESA, 1993), 28–34. Michelangelo De Maria, *The History of ELDO Part 1: 1961–1964,* HSR-10 (Noordwijk, The Netherlands: ESA, 1993), 7–12, 17. *CECLES-ELDO 1960–1965,* report to the Council of Europe, Paris, 23 December 1965, 7–16.

20. *ELDO Convention,* art. 6(1), 6(2), 9, 11, 16(1), and 16(2). Memorandum from Arnold Frutkin to Robert F. Packard, Director Outer Space Affairs, SCI, Department of State, Subject: Embassy Brussels observations on ELDO, 5 July 1966, NASAHO.

21. MAU = million accounting units, a unit calculated from a mix of European currencies. In the mid-1960s, one AU was approximately equivalent to one dollar.

22. De Maria, *History of ELDO Part 1,* 17–19, 23, 24, 27, 30. Krige and Russo, *Europe in Space,* 71. *CECLES-ELDO 1960–1965,* 15–17.

23. *CECLES-ELDO 1960–1965,* 18–20. Andrew J. Butrica, ed., *Beyond the Ionosphere: Fifty Years of Satellite Communication,* SP-4217 (Washington, D.C.: NASA, 1997).

24. Krige and Russo, *Europe in Space,* 71–82. Michelangelo De Maria and John Krige, "Early European Attempts in Launcher Technology: Original Sins in ELDO's Sad Parable," *History and Technology* 9 (1992): 109–37.

25. Quotation from "Intra-European Cooperation on Space Programs," author probably Frutkin, ca. 1968, NASAHO. Lorenza Sebesta, *The Availability of American Launchers and Europe's Decision 'To Go It Alone',* HSR-18 (Noordwijk, The Netherlands: ESA, 1996).

26. *World-Wide Space Activities,* Report Prepared for the Subcommittee on Space Science and Applications of the Committee on Science and Technology, U.S. House of Representatives, 95th Congress, 1st Session, by the Science Policy Research Division of the Congressional Research Service, September 1977, 263, 266.

27. Werner Büdeler, *Raumfahrt in Deutschland: Forschung, Entwicklung, Ziele* (Frankfurt: Econ Verlag, 1978), 31–37, 141–43. Peter Fischer, *The Origins of the Federal Republic of Germany's Space Policy 1959–1965—European and National Dimensions,* HSR-12 (Noodwijk, The Netherlands: ESA, 1994), 35–37, 41–45, table II.

28. Groupe ad hoc de la Gestion du Controle, "Note de la Délégation belge," ELDO/C(66)26, 3 mars 1966, Paris, HAEUI; *Organisation Europeenne pour la mise au point et la construction de lanceurs d'engins spatiaux, structure et table des matieres du rapport de synthese de la commission de revue de projet, 1ère partie, Rapport de Synthèse de la Commission de Revue de Projet,* Neuilly, 19 avril 1972, ELDO/CRP(72)38, HAEUI Fond 2885, 8.

29. *European Space Vehicle Launcher Development Organisation, Europa II Project Review Commission, Group No. 5, Sequencing, Separation and Safety Systems Final Report,* Neuilly, 18 April 1972, ELDO/CRP(72)40, HAEUI Fond 2887, Section 0.4.7.1. *Rapport de Synthèse, 1ère partie,* 6–7.

30. *Final Report of the Project Review Commission,* ELDO/C(72)18, Neuilly, France, 19 May 1972, HAEUI, 14.

31. *Europa II Project Review Commission, Group No. 5,* sections 0.4.1.1, 0.4.2.3.

32. *Final Report of the Project Review Commission,* 13, 14, 17.

33. *Final Report of the Project Review Commission,* 12–13. *Rapport de Synthèse, 1ère partie,* 8.

34. *Europa II Project Review Commission, Group No. 5,* section 0.4.2.1. *Final Report of the Project Review Commission,* 13.

35. *European Space Vehicle Launcher Development Organisation, Europa II Project Review Commission, Group 3 Final Report,* Neuilly, 18 April 1972, ELDO/CRP(72)40, HAEUI Fond 2884, 15–18; quotation at 15. *Europa II Project Review Commission, Group No. 5,* section 0.4.6.2.

36. J. Nouaille, "The ELDO PAS Programme and Its Management," *ESRO/ELDO Bulletin* 1 (May 1968): 10. I. Stevenson, "L'analyse de Réseau, contribution à la gestion du CECLES," *ESRO/ELDO Bulletin* 2 (August 1968): 12–14.

37. Stevenson, "L'analyse de Réseau," 15.

38. *CECLES-ELDO 1960–1965,* 61–65.

39. L. T. D. Williams, Chairman, Corps of Inspectors, letter to Secretary-General, ELDO/CECLES, ELDO/T(67)5, London, 17 December 1966, HAEUI ELDO Fond 2958. See Annex B, which includes ELDO/PG(63)T/27, "Report on ELDO Initial Programme, October 1963—Part III—Division of Responsibilities," and "Technical Acceptance Procedures for Application to F4 Vehicle."

40. Williams, letter to Secretary-General, 17 December 1966.

41. Groupe ad hoc de la Gestion du Controle, "Note de la délégation belge," ELDO/C(66)26, 3 mars 1966, Paris, HAEUI.

42. Nouaille, "ELDO PAS Programme," 10.

43. "14 Session of ELDO Council," Paris, 3 March 1966, Discussion of "Structure of Secretariat," 15–19, HAEUI ELDO 1140. "15 Session of ELDO Council," 30 March 1966, Paris, Section on "Management and Control of ELDO Programmes," 7, 8, HAEUI ELDO 1141. Quotations from "Decision of the Council on the ELDO Structure 1966," European Space Vehicle Launcher Development Organisation, ELDO/C(66)27, Paris, 12 April 1966, HAEUI ELDO 1177.

44. "Resolution of the Council on the Management and Control Procedures for the Programmes," European Space Vehicle Launcher Development Organisation, ELDO/C(66)27, Paris, 12 April 1966, HAEUI ELDO 1177.

45. Europa II was also known as the ELDO-PAS (Perigee-Apogee System).

46. "Creation of an ELDO Integrating Group," 2 December 1966, ELDO Fond 1210, HAEUI; "21 Session of ELDO Council," Paris, 15 and 16 December 1966, section on "Interim Report by the Secretariat on the constitution of the Industrial Integrating Group," ELDO/C(66)64, 16–18, HAEUI ELDO 1147. "Industrial Integrating Group," Note by the Secretariat, ELDO/T/(67)1, Paris, 24 January 1967, HAEUI ELDO Fond 2954.

47. Nouaille, "ELDO PAS Programme," 10–11. *World-Wide Space Activities,* 271–72; "17 Session of ELDO Council," Paris, 8 July 1966, Section on "Modifications to Structure," 2, 3, HAEUI ELDO 1143. "Draft Resolution-Integrating Group," ELDO/C(66)WP/16, Paris, 29 September 1966, HAEUI ELDO 1229. "ELDO Integrating Group Operating Proposal in SEREB International Division," 28 November 1966, HAEUI ELDO 1210. Krige and Russo, *Europe in Space,* 76.

48. "Industrial Integrating Group," Note by the Secretariat, ELDO/T/(67)1, Paris, 24 January 1967, HAEUI ELDO Fond 2954.

49. Williams, letter to Secretary-General, 17 December 1966.

50. "Comments by the Secretariat on interface problems following the first report of the Corps of Inspectors," Note by the Secretariat, ELDO/T(67)5, Paris, 7 February 1967, HAEUI Fond 9458.

51. Sebesta, *Europe's Decision 'To Go It Alone,'* 20–21. *Control and Management of ELDO Programmes,* Report by the Secretary-General, ELDO/C(67)47, 9 October 1967, ELDO Fond 1288, HAEUI.

52. T. A. Graham and E. Dombrowski, "Electrical Interface Problems in Multi-Stage Launchers as Illustrated in the ELDO Europa Project," *ESRO/ELDO Bulletin* 12 (November 1970): 16–23.

53. "Report on Launch of F6/1 Vehicle," Note by the Secretariat, ELDO/T(67)24, Paris, 22 September 1967, HAEUI ELDO Fond 2977.

54. "Report on Launch of F6/2 Vehicle," Note by the Secretariat, ELDO/T(68)1, Neuilly, 8 February 1968, HAEUI ELDO Fond 2992.

55. "Consequences of F6/2 on the Programme," Note by the Secretariat, ELDO/T(68)2, Neuilly, 20 February 1968, HAEUI ELDO Fond 2993

56. "Operation Coralie Development Plan," Annex A to ELDO/T(68)2, Courbevoie, 31 January 1968, SEREB Technical Directorate, DTA/C-no. 30/9497, HAEUI ELDO Fond 2993.

57. "Secretariat Comments on the French Authorities' Proposal," Annex B to ELDO/T(68)2, 20 February 1968, HAEUI ELDO Fond 2993.

58. Quotations from "Functioning of the Project Management Directorates Created in the Secretariat," Note by the Secretariat, ELDO/C(68)12, Neuilly, 9 May 1968, HAEUI. Minutes of the 29th Session of the Council held in Neuilly on 24 and 25 June 1968, ELDO/C(68)12 (and corrigendum), HAEUI 1308.

59. P. Rochefort, "The Flight Test of the Europa I Satellite Launcher," *ESRO/ELDO Bulletin* 4 (January 1969): 14–19. "F7 Flight Test," Note by the Scientific and Technical Committee," ELDO/T(68)20, Neuilly, 11 December 1968, HAEUI ELDO Fond 3010.

60. "Preliminary Report on the Launch of the F8 Vehicle," Note by the Secretariat, ELDO/T(69)8, Neuilly, 17 September 1969, HAEUI ELDO Fond 3026.

61. "F9 Assessment-Progress Report," Note by the Secretariat, ELDO/T(70)10, Neuilly, 9 October 1970, HAEUI Fond 3052. *Final Report of the Project Review Commission,* 16.

62. *Final Report of the Project Review Commission,* 29–30. R. Aubinière, "Tenth Anniversary of the Establishment of ELDO," *ESRO/ELDO Bulletin* 24 (March 1974): 13.

63. *Final Report of the Project Review Commission,* annex 2.

64. *Final Report of the Project Review Commission,* 15, 20–21; quotation at 20. U.S. Department of Commerce, Bureau of International Commerce, *Report of the Space Industry Trade Mission on the European Aerospace Industry,* January 1966, "U.S.-Europe 1965–1972" folder 014548, LEK 7/9/4, NASAHO.

65. *Preliminary Report of the Europa II Project Review Commission to the Council of ELDO,* annex V, 9–10, in *Final Report of the Project Review Commission;* "Europa III, Preparatory Phase," *ESRO/ELDO Bulletin* 18 (May 1972): 10–14. ELDO adopted these American methods for *Europa III* planning even before the F11 failure. It could do this because the program was in the planning stages only, not an officially approved program subject to the national governments.

66. *World-Wide Space Activities,* 282–87.

Seven | ESRO's American Bridge across the Management Gap

J. J. Beattie and J. de la Cruz, "ESRO and the European Space Industry," *ESRO Bulletin* 3 (1967): 4.

1. A version of this chapter appears in Stephen B. Johnson, "Building an American Bridge over the 'Management Gap': The Adoption of Systems Management in ESRO and ESA," *History and Technology* 16 (1999): 1–32.

2. Michelangelo De Maria, *Europe in Space: Edoardo Amaldi and the Inception of ESRO,* HSR-5 (Noordwijk, The Netherlands: ESA, 1993), 1–9; quotations at 8–9.

3. De Maria, *Edoardo Amaldi,* 9–14.

4. De Maria, *Edoardo Amaldi,* 14–18. John Krige, *The Prehistory of ESRO,* HSR-1 (Noordwijk, The Netherlands: ESA, 1992), 2–5. Most of these scientists and administrators were acquaintances of Amaldi and Auger through CERN.

5. Krige, *Prehistory of ESRO,* 5–17. De Maria, *Edoardo Amaldi,* 22–28.

6. Krige, *Prehistory of ESRO,* 17–27.

7. John Krige, *The Early Activities of the COPERS and the Drafting of the ESRO Convention (1961/62),* HSR-4 (Noordwijk, The Netherlands: ESA, 1993), 1–29. G. Phélizon, "Management of ESRO I Satellite Program," *Electrical Communication* 44, no. 1 (1969): 60.

8. Krige, *Early Activities of the COPERS,* 29–44. John Krige, *Europe into Space:*

The Auger Years (1959–1967), HSR-8 (Noordwijk, The Netherlands: ESA, 1993), 23–28. Against their will, the British had to support substantial cost increases at CERN when outvoted in the Council.

9. "The Operation and Structure of ESRO," *ESRO Bulletin* 1-2 (May 1966): 2–4.

10. Beattie and de la Cruz, "ESRO and the European Space Industry," 3, 4.

11. Krige, *Auger Years,* 50–53. Arturo Russo, *ESRO's First Scientific Satellite Programme 1961–1966,* HSR-2 (Noordwijk, The Netherlands: ESA, 1992) and *Choosing ESRO's First Scientific Satellites,* HSR-3 (Noordwijk, The Netherlands: ESA, 1992).

12. Contracts went to Contraves A.G. of Zurich, Switzerland, and Bell Telephone Manufacturing Company of Antwerp, Belgium.

13. See "ESRO-Ib," ESRO, n.d., ca. 1969, ESTEC, 19. Manfred G. Grensemann, "Systems Engineering and Project Management of the ESRO I Satellite," in *Project Management and Project Control, 10th ESRO Summer School,* Frascati, September 1972, ESTEC, 119. M. G. Grensemann, "ESRO-I Project Management and Organization," *ESRO/ELDO Bulletin* Suppl. (April 1969): 27.

14. Ateliers de Constructions Electriques of Charleroi, Belgium, and Eidgenössische Technische Hochschule of Zurich, Switzerland.

15. Subcontracts went to Société Engins Matra, Sperry, and Ferranti.

16. Ants Kutzer, "The Preliminary Design of the ESRO II Satellite," ESRO ESTEC, August 1964, ESTEC, 4–7. "The ESRO-II Satellite," *ESRO Bulletin* 1 (1967): 2–5 (author "R. J. D."); quotation at 5.

17. Junkers teamed with British Aircraft Corporation, Études Techniques et Constructions Aerospatiales of Belgium, Messerschmitt AG, and Société Nationale d'Études et de Constructions de Moteurs d'Aviation of Paris.

18. *HEOS A,* European Space Research Organisation, ca. November 1968, ESTEC, 1–4. J. A. Vandenkerckhove, "Reflexions sur la Gestion du Projet HEOS-1," *ESRO/ELDO Bulletin* Suppl. HEOS-1 (August 1969): 6–13. Eberhard Wunderer, "HEOS-1, Ein ESRO—Satellit unter Deutscher Projektleitung," *DGLR Jahrbuch* (1968): 153–55.

19. "Memorandum of Understanding between the European Space Research Organisation and the United States National Aeronautics and Space Administration on the preparation, launching, and use of ESRO-I and ESRO-II satellites," 8 July 1964, reprinted in Lorenza Sebesta, *United States-European Cooperation in Space during the Sixties,* HSR-14 (Noordwijk, The Netherlands: ESA, 1994), app. 5. A. Kutzer, "The ESRO II Satellite Project," *18th International Astronautical Congress* 2 (1967): 153. Wolfgang Nellessen, "Design and Construction of ESRO II," paper presented at Seminar ES 2 for the course entitled Environments and Their Role in Spacecraft Technology, ESRO Summer School, Noordwijk, The Netherlands, August 1968, ESTEC, 6–8. I have not been able to determine if ESRO personnel went to GSFC for training. NASA's offer extended also to national programs, including building and launching British satellites, exchanging launches with France, launching a German-built satellite and later jointly developing a solar probe, launching Italian satellites with American launchers from American and Italian sites, and agreeing to launch Spanish and Swedish satellites. French and Italian engineers also visited Goddard for a

year to learn about satellite technologies and management techniques. John Krige and Lorenza Sebesta, "US-European Co-operation in Space in the Decade after Sputnik," in Giuliana Gemelli, ed., *Big Culture, Intellectual Cooperation in Large-Scale Cultural and Technical Systems: An Historical Approach* (Bologna: Editrice, 1994), 275–78. "Notes on ESRO Situation," 18 September 1968, authorship unknown, probably by NASA International Coordinator Frutkin, NASAHO. Claude Carlier and Marcel Gilli, *The First Thirty Years at CNES, The French Space Agency, 1962–1992* (Paris: La Documentation Française/CNES, 1994), 18.

20. L. A. Husain, "Notes on a US Visit, October 19th to October 23rd, 1964," ESRO EWP-55, LSD/WP 7 ESTEC. Alfred Rosenthal, *Venture Into Space: Early Years of Goddard Space Flight Center,* SP-4301 (Washington, D.C.: NASA, 1970), 43–49.

21. "NASA-ESRO ESRO I Satellite 2nd Joint Working Group Meeting," Delft, 20–22 September 1965, HAEUI Fond 50147. "NASA ESRO ESRO II Satellite 3rd Joint Working Group Meeting," Delft, 20 September, Stevenage, 23 September, Paris, 24 September 1965, HAEUI Fond 50147. Nellessen, "Design and Construction of ESRO II," 12–13. Grensemann, "Systems Engineering and Project Management," 128.

22. "Europeans Getting Space Tips at Rice," *Houston Post,* 2 February 1965.

23. C. R. Hume, "Project Management and Technical Progress of the ESRO II Satellite Project," *Aeronautical Journal of the Royal Aeronautical Society* 72 (January 1968): 95. Wunderer, "HEOS-1," 153. "Junkers Signs Satellite Contract with Lockheed," Released by Newsbureau, Lockheed Missiles & Space Company, Sunnyvale, Calif., Friday, 3 June 1966, NASAHO. "Joint Ventures Between US and European Firms in the Space Technology Area," author probably Frutkin, ca. 1968, NASAHO.

24. L. A. Potter, "Notes on an Informal Visit to Ministry of Aviation London on November 20th, 1964," ESRO EWP-59, LSD/WP 11, 4 December 1964, ESTEC.

25. Ants Kutzer, "Implementation of a Satellite Project," paper presented on the occasion of leaving ESTEC on 15 September 1967, HAEUI Fond 51048, 14–17; quotation at 17.

26. Ibid., 5–10; quotation at 9. Merton J. Peck and Frederic M. Scherer, *The Weapon Acquisition Process: An Economic Analysis* (Cambridge: Division of Research, Graduate School of Business Administration, Harvard University, 1962). A. W. Marshall and W. H. Meckling, "Predictability of the Cost, Time and Success of Development," Report P-1821 (Santa Monica, Calif.: RAND Corporation, December 1959).

27. Kutzer, "Implementation of a Satellite Project," 15–17; quotation at 16. Kutzer, "ESRO II Satellite Project," 161. Grensemann, "Systems Engineering and Project Management," 126.

28. J. Henrici, "Das Projekt-Management HEOS-A als Beispiel für die Auslegung und Durchführung eines europäischen Satelliten," Symposium über Projekt-Management, Bad Godesberg, September 1967, 135. The author's translation. The German text reads as follows:

> Die Firmen hatten ihre Vorschläge für das Angebot in Europa gegenseitiger Abstimmung gemacht und trafen sich anschließend in Sunnyvale zur Anfertigung

des endgültigen Angebottextes. In diesen Wochen der sehr lebhaften Diskussion mit den erfahrenen Spezialisten der amerikanischen Firma entwickelte sich ein starker Kontakt zwischen den Führungskräften der europäischen Firmen der für die Zusammenarbeit bei der Verwirklichung des Projektes entscheidend wurde. Zudem lernten die Beteiligten mit den gleichen Worten die gleichen Vorstellungen zu verbinden . . . Das Angebot der Firmengruppe umfaßte etwa 1,000 Seiten, von denen rund ein Drittel reine Fragen des Management und der Kostenverteilung behandelten. Insbesondere dieser Teil hätte ohne die Beratung der amerikanischen Firma nicht eine so tiefe Bearbeitung erfahrhen.

29. Wunderer, "HEOS-1," 153, author's translation. The German reads, "Ein erheblicher Teil der Entwicklungsarbeiten war bereits vorweggenommen, und es konnte mit großer Sicherheit angenommen werden, daß der Anbieter darauf aufbauend eine ebenso reale Zeit/Kostenplanung estellt hatte."

30. Phélizon, "Management of ESRO I Satellite Program," 55. Grensemann, "Systems Engineering and Project Management," 123–27.

31. B. J. Madauß, "Erfahrungen mit dem PERT-Planungsverfahren beim Satellitenprojekt HEOS-A," *Luftfahrttechnik-Raumfahrttechnik* 14, no. 7/8 (1968): 174–80.

32. Hume, "Project Management and Technical Progress," 96. Kutzer, "Implementation of a Satellite Project," 18–20; quotation at 20.

33. Hume, "Project Management and Technical Progress," 96.

34. Kutzer, "Implementation of a Satellite Project," 22–24. Hume, "Project Management and Technical Progress," 97.

35. "Environmental Testing at ESTEC," *ESRO Bulletin* 1–2 (1966): 11–12. "Brief Summary of Major Environmental Test and Research Facilities," *ESRO Bulletin,* Mars 1968, 12–14. "News in Brief," *ESRO Bulletin* 3 (1966): 10. H. Busch, "The Work of the Spacecraft Projects Department at ESTEC," *ESRO Bulletin* 4 (1966): 13, 15. "Progress in Operational Programmes," *ESRO Bulletin* 4 (1966): 17. "ESRO-II Satellite," 15 (author: "R. J. D."). Phélizon, "Management of ESRO I Satellite Program," 55–56.

36. Electronic components include transistors, amplifiers, and resistors.

37. *HEOS A,* 7. Vandenkerckhove, "Reflexions sur la Gestion du Projet HEOS-1," 10. Grensemann, "Systems Engineering and Project Management," 121. Hume, "Project Management and Technical Progress," 100. R. W. Young and A. J. Clarke, "Reliability Aspects of the ESRO-II Satellite Programme," paper presented at the IEE Conference on Reliability in Electronics, 10–12 December 1969, 13. I do not know why *ESRO-II* escaped this problem but *ESRO-I* and *HEOS* did not. Space-qualified components were under the control and prioritization of the U.S. Department of Defense.

38. Phélizon, "Management of ESRO I Satellite Program," 59.

39. Grensemann, "Systems Engineering and Project Management," 122. Grensemann, "ESRO-I Project Management and Organization," 23–28. "News in Brief," *ESRO Bulletin* 3 (1966): 10. "Progress in Operational Programmes," *ESRO Bulletin* 4 (1966): 16–17. "Progress in Operational Programme," *ESRO Bulletin* 5/6 (1966): 12. "Progress in Operational Programme," *ESRO Bulletin* (March 1968): 19. *HEOS A,* 7.

40. Young and Clarke, "Reliability Aspects," 12. D. Lennertz, "The HEOS-1 Launch Operations," *ESRO/ELDO Bulletin,* Suppl. HEOS-1 (August 1969): 37–43. Grensemann, "ESRO-I Project Management and Organization," 28.

41. Krige, *Auger Years,* 58–61. Ironically, the British had insisted on the stringent financial controls, and these controls led to cancellation of the Large Astronomical Satellite, Britain's primary scientific interest. This, along with the experience of ELDO, made the British skeptical of cooperative European efforts in the late 1960s.

42. Krige, *Auger Years,* 58–61.

43. Vandenkerckhove, "Reflexions sur la Gestion du Projet HEOS-1," 10–13. Kutzer, "Implementation of a Satellite Project," table 1.

44. "The TD-1, TD-2 Satellites," *ESRO Bulletin* 2 (1967): 5–6. Krige, *Auger Years,* 61.

45. Similar problems with three-axis stabilization led to many of JPL's problems with the *Ranger* design in 1962–64.

46. "TD Project," *ESRO/ELDO Bulletin* 2 (August 1968): 39. "TD Project," *ESRO/ELDO Bulletin* 3 (November 1968): 41–42.

47. Quotation from Arturo Russo, *The Early Development of the Telecommunications Satellite Programme in ESRO,* HSR-9 (Noordwijk, The Netherlands: ESA, 1993), 42. Andrew J. Butrica, ed., *Beyond the Ionosphere: Fifty Years of Satellite Communication,* SP-4217 (Washington, D.C.: NASA, 1997).

48. *Satellites Européens de Telecommunications, Addenda 3, Maitrise d'Oeuvre du Projet CETS,* Mai 1968, Organisation Européene de Recherches Spatiales, HAEUI Fond 51048. Author's translation. The French states: "Quoique les avantages de la proposition du CERS aient été reconnus, en particulier en ce qui concerne la flexibilité du choix des contractants, le contrôle du coût du programme et la répartition géographique des tâches, certaines délégations ont exprimé très clairment l'avis que l'industrie devait se voir confier la responsibilité globale du système, car cette tâche lui permettrait d'acquérir une expérience hutement profitable dans le domaine de la gestion technique et financière de projets complexes."

49. Krige, *Auger Years,* 58.

50. *Report of the Group of Experts to Study the Internal Structure, Procedures, and Methods of ESRO* (Bannier Report), 29 March 1967, ESRO/C/APP/48, HAEUI, 53–54, and chap. 4; quotation at 54. Krige, *Auger Years,* 35–47.

51. Quotation from M. Schalin, "Improvement of Management Procedures for Major Satellite Projects," GEN/WP/98, Neuilly, 26 March 1968, HAEUI Fiche M28, 1. "Management of Satellite Projects," Interim Report by the Director General, ESRO, ESRO/C/343, Att: GEN/WP/98, 26 March 1968, HAEUI Fiche M28.

52. Schalin, "Improvement of Management Procedures," 2–3. H. Marin, "A Suggested Configuration for Working Groups on New Projects of Medium Scientific Satellites," ESTEC Internal Working Paper No. 118, November 1967, ESRO, ESTEC. J. F. Lafay, "Utilisation des formuls Goddard et IITRI pour l'évaluation du coût des satellites, Application aux programmes ESRO," Département des Satellites et Fusées Sondes, European Space Research and Technology Centre, Internal Working Paper No. 277, 27 Juin 1968, ESTEC.

53. Schalin, "Improvement of Management Procedures," 4–8.

54. ESRO Council, ESRO/C/367, rev. 1, fall 1968, HAEUI Fiche M32, 14–16. Initially there were two projects, TD-1 and TD-2. Because of cost cuts, these were compressed to a single combined project, TD-1/2, and finally revised and renamed TD-1.

55. European Space Research Organisation Council, "The TD-1 Special Project," ESRO/C/362, Att.: Annexes I to VI, Neuilly, 23 September 1968, HAEUI Fond 5874, 8–14; quotation at 8. For comparison, ESRO-I used only nineteen project personnel and was considered a large project in 1966.

56. European Space Research Organisation Council, "The TD-1 Special Project," ESRO/C/362, Att.: Annexes I to VI, Neuilly, 23 September 1968, HAEUI Fond 5874, 8–14; quotation at 14. "ESRO News," *ESRO/ELDO Bulletin* 6 (July 1969): 26–28.

57. ESRO Council, *Phased Planning for Scientific Satellite Projects Guidelines,* ESRO/C/405, Att.: DG/PS/6137, Neuilly, 19 June 1969, HAEUI Fiche M37. ESRO Scientific and Technical Committee, *Procedures for the Award of Major Satellite Contracts,* ESRO/ST/326, Neuilly, 13 October 1969, HAEUI 9808.

58. "First Interim Report on the Implementation of an M.I.S. in ESRO," Management Information Systems Study Group, WO-522 Rev. 1, January 1969, HAEUI Fond 51083, app. 1, 1. For ESTEC, see prior references to the work of H. Marin and project management improvements for TD-1.

59. "First Interim Report on the Implementation of an M.I.S. in ESRO," 6–8.

60. Memorandum from H. Hoernke to members of the MISS Group, Subject: Report on management information in project control, 20 November 1968, ESTEC/PC/HH/dp/18.192/34, HAEUI Fond 1088 (emphasis in original).

61. I did not find what PMS stands for, but it is undoubtedly "Project Management System." North American Rockwell had started working with IBM on this system in the mid-1960s based on North American Rockwell's experience on *Apollo.*

62. Gehriger, "Management Services at ESTEC," n.d. but soon after 27 November 1968, HAEUI Fond 51083.

63. These are all debatable points, but this is what they believed.

64. "First Interim Report on the Implementation of an M.I.S. in ESRO," 9–11.

65. "First Interim Report on the Implementation of an M.I.S. in ESRO," 17–19.

66. A. Dattner, "Development of a Management Information System at the European Space Research Organisation," *Journees D'Electronique, Colloque Internationale,* 2–5 Mars 1971, Universite de Toulouse, HAEUI Fond 51351.

67. Hellmuth Gehriger, "The ESTEC Project Control System," *ESRO/ELDO Bulletin* 19 (July 1972): 22–32.

68. Memorandum DOP/86 from N. Longdon, Head of Personnel, to R. Gibson, Subject: Organisation of satellites work at ESTEC, 9 January 1970, HAEUI Fond 51928. Quotations from Memorandum G/2197/sj from R. Gibson to Blassel, Subject: Organisation of satellite work at ESTEC, 23 February 1970, HAEUI Fond 51928.

69. U. Montalenti, "Role of ESOC in Management of the Application Programme (Note for the Directorate)," 14 December 1971, HAEUI Fond 51928. J. B. Lagarde, Chef

de la Division Études de Systèmes, note pour M. Blassel, Departement Satellites & Fusees-Sondes, SYST/1800/JBLlv, 13 janvier 1972, Objet: Relations ESTEC/ESOC concernant METEOSAT, HAEUI Fond 51928. Memorandum from O. Hammarström, Director, ESTEC, to Director, ESOC, Subject: Integrated project management, 11 February 1972, HAEUI Fond 51928. A. Vandormael, Chef du Département A&F, to Chef Division Contrats, Chef du Personnel, Chef Division Finance, Chef Administrative Services Group, 8 decembre 1972, Objet: Support to projects, HAEUI Fond 51928. Memorandum from A. Vandormael to Director, ESTEC, Objet: Projet DII "Support to Projects," 5 janvier 1973, HAEUI Fond 51928. J. Toussaint, TEI/DIR/6564/JT/WJ, 29 January 1973, Project Support Activities—Definition of "Project Support" and "Common Support," HAEUI Fond 51928. O. Hammarström, "Support to Projects," EII 1/73, 14.2.73, HAEUI Fond 51928.

70. John Krige and Arturo Russo, *Europe in Space 1960–1973,* SP-1172 (Noordwijk, The Netherlands: ESA, 1994), 109–16. Théo Lefèvre, "European Space Conference, Brussels 1972," *ESRO/ELDO Bulletin* 20 (February 1973): 5–7. "European Space Conference," *ESRO/ELDO Bulletin* 22 (August 1973): 2–4. H. Kaltenecker, "La nouvelle agence spatiale europeenne," *ESRO/ELDO Bulletin* 26 (December 1974): 16–18, 31–33. John Krige, "The European Space System," in Krige and Arturo Russo, *Reflections on Europe in Space,* HSR-11 (Noordwijk, The Netherlands: ESA, 1994), 1–11.

71. ESRO, Ariane Launcher Interim Board, "Draft Agreement between the European Space Research Organisation and the Centre National D'Études Spatiales (France) Concerning the Execution of the Ariane Launcher Programme," ESRO/IB-LIIIS(73)3, Att: ESRO/C(73)41, Annex II, rev. 1, Neuilly, 5 November 1973, HAEUI Fond ESRO 8807. Quotation from ESRO, Ariane Launcher Programme Board, "Report on the Ariane Programme Definition Stage Commentary by the Secretariat," ESRO/PB-ARIANE(74)9, Neuilly, 15 February 1974, 9. "Ariane Launcher," *ESRO/ ELDO Bulletin* 23 (November 1973): 20–21. R. Vignelles and P. Rasse, "The Ariane Launcher and Its Progress," *ESA Bulletin* 15 (August 1978): 10–21. P. Gauge, "Stage-Separation Testing for Ariane," *ESA Bulletin* 15 (August 1978): 67–70. Emmanuel Chadeau, ed., *L'Ambition Technologique: Naissance d'Ariane* (Paris: Institute d-Histoire de l'Industrie, 1995).

72. ESRO, ARIANE Launcher Interim Board, "Draft Agreement between the European Space Research Organisation and the Centre National D'Études Spatiales (France) Concerning the Execution of the ARIANE Launcher Programme," ESRO/IB-LIIIS(73)3, Att: ESRO/C(73)41, Annex II, rev. 1, Neuilly, 5 November 1973, HAEUI Fond ESRO 8807. ESRO, ARIANE Launcher Programme Board, "Report on the Ariane Programme Definition Stage Commentary by the Secretariat," ESRO/PB-ARIANE(74)9, Neuilly, 15 February 1974. "Ariane Launcher," *ESRO/ELDO Bulletin* 23 (November 1973): 20–21. R. Vignelles and P. Rasse, "The Ariane Launcher and Its Progress," *ESA Bulletin* 15 (August 1978): 10–21. Emmanuel Chadeau, ed., *L'ambition technologique: Naissance d'Ariane* (Paris: Institute d'Histoire de l'Industrie, 1995).

73. On German motivations, see comments by H. A. Strub in *Proceedings of the*

Workshop on the History of Spacelab, ESTEC, Noordwijk, The Netherlands, 22–23 April 1997, SP-411 (Noordwijk, The Netherlands: ESA, 1997), 36. "Distribution of Responsibilities between Head of Spacelab Programme and Director of ESTEC," Spacelab, Draft 28.6.1973, HAEUI 50975. Memorandum D/559/6931 from O. Hammarström, Director of ESTEC, to Director of Administration, ESRO, Subject: Spacelab Programme-Organigram, 18 October 1973, HAEUI 50975. "Briefings on 'Skylab lessons learned' by NASA," ESRO ESTEC Internal Memorandum LSO/088/LE/mb, 4 September 1974, EUI Archive File 60772. *World-Wide Space Activities,* Report Prepared for the Subcommittee on Space Science and Applications of the Committee on Science and Technology, U.S. House of Representatives, 95th Congress, 1st Session, by the Science Policy Research Division of the Congressional Research Service, September 1977, 67. McDonnell Douglas provided thirty-five consultants to ERNO, five to Aeritalia, and two to Fokker. See Lorenza Sebesta, *Spacelab in Context,* HSR-21 (Noordwijk, The Netherlands: ESA, 1997), 23.

74. Sebesta, *Spacelab in Context,* 25, 28–29. *World-Wide Space Activities,* 68–71. Arturo Russo, *Big Technology, Little Science: The European Use of Spacelab,* HSR-19 (Noodwijk, The Netherlands: ESA, 1997): 10–11.

75. H. Stoewer, in *Workshop on the History of Spacelab,* 97.

76. ESRO became ESA at this time.

77. Douglas R. Lord, *Spacelab: An International Success Story,* SP-487 (Washington, D.C.: NASA, 1987), 86, 108. Comments by K. Berge and H. Stoewer, in *Workshop on the History of Spacelab,* 96–97.

78. "Standardisation in the field of computer software," ESTEC Internal Memorandum TI/MT/dc/7288 by M. Trella, 4 May 1976, EUI Archive File 60567. "European Space Agency, Spacelab Programme Board, Sixth Meeting Draft Minutes, Meeting held at Neuilly on 28 September 1976," ESA/PB-SL/MIN/6, Paris, 25 October 1976, HAEUI. "Director General's Instruction Concerning the Monitoring of Contract Actions for Software Procurement by the Board for Software Standardisation and Control," Draft Memorandum from the ESA Director General's Office, date unreadable, Paris, HAEUI 60432. "Standardized Development of Computer Software," class held by Robert C. Tausworthe of JPL, on 13–15 November 1978, Doc. No. BSSC-INF 1, 12 October 1978, HAEUI 60567. Lord, *Spacelab,* 206–209. Comments by H. E. W. Hoffman, *Workshop on the History of Spacelab,* 90.

79. Sebesta, *Spacelab in Context,* 27. Comments by B. Pfeiffer, *Workshop on the History of Spacelab,* 85.

80. Sebesta, *Spacelab in Context,* 29. Quotation from Comments by B. Pfeiffer, 85.

81. Sebesta, *Spacelab in Context,* 22, 30.

82. Ibid., 31–34.

83. H. Stoewer, "The Systems Engineering Department at ESTEC," *ESA Bulletin* 19 (August 1979): 66–69; quotation at 66.

Eight | Coordination and Control of High-Tech Research and Development

1. One might also classify Pickering as an entrepreneur, as he led JPL into full-scale development on *Corporal* and then into the space program. Jack James remembers him as an excellent leader but not a good manager.

2. Sidney Siegel, "NASA's Configuration Management System and Conflict Management within the Internal Organization of the Urban System," in S. R. Siegel, *Management Technology Applied to Urban Systems* (Camden, N.J.: Rutgers University, 1972), 69–76.

3. Michael E. Brown, *Flying Blind: The Politics of the U.S. Strategic Bomber Program* (Ithaca, N.Y.: Cornell University Press, 1992), chaps. 1 and 9. H. E. McCurdy, "The Cost of Spaceflight," *Space Policy* 11, no. 2 (1995): 144–49.

4. Corporal went from 45–50% to 60–70%, and Atlas and Titan improved from 50% to the 80% range.

5. JPL implemented systems management between *Ranger 5* and *Ranger 6*. *Ranger 6* failed, but the next three flights succeeded.

6. This is another way of stating Max Weber's idea that bureaucracies adopt rules to achieve reliability. In "Three Approaches to Big Technology: Operations Research, Systems Engineering, and Project Management," *Technology and Culture* 38, no. 4 (October 1997): 891–919, I called this "procedural knowledge." Since that time, I noted another literature in economics that has identified this as "codification," or "codified knowledge." See R. Cowan and D. Foray, "The Economics of Codification and the Diffusion of Knowledge," *Industrial and Corporate Change* 6, no. 3 (1997): 595–622.

7. Hendrick Bode, "The Systems Approach," in *Applied Science and Technological Progress,* A Report to the Committee on Science and Astronautics, U.S. House of Representatives by the National Academy of Science (June 1967), 73–94.

8. On systematic management, see JoAnne Yates, *Control Through Communication: The Rise of System in American Management* (Baltimore: Johns Hopkins University Press, 1989). On scientific management, see Hugh G. H. Aitken, *Taylorism at Watertown Arsenal: Scientific Management in Action, 1908–1915* (Cambridge: Harvard University Press, 1960), and Daniel Nelson, ed., *A Mental Revolution: Scientific Management since Taylor* (Columbus: Ohio State University Press, 1992).

9. The *Challenger* accident occurred during operations, not during R&D. Systems management, an R&D management scheme, perhaps should never have been applied for operations, but it was the only system NASA knew. Systems management contributed to this disaster not because of its bureaucracy per se but because through systems management, managers gained control over engineers. Managers then overruled engineering concerns to keep to an operational schedule. On *Hubble,* to save costs, managers eliminated tests that would have revealed *Hubble*'s myopia. Is "the system," namely the bureaucratic procedures, to blame for these failings? As a communication and control system, systems management revealed the problem on *Challenger* and was

reduced on *Hubble* so that it could not find the problem. For both projects, the issues are more complex than a simple "systems management is or is not to blame" stance. Both involve systems management, political pressures, relationships between industry and government, and simple miscommunications and errors.

10. William Minoru Tsutsui, *Manufacturing Ideology: Scientific Management in Twentieth-Century Japan* (Princeton, N.J.: Princeton University Press, 1998).

11. Ida R. Hoos, *Systems Analysis in Public Policy: A Critique* (Berkeley: University of California Press, 1972).

12. See Stephen B. Johnson, *The United States Air Force and the Culture of Innovation, 1945–1965* (Washington, D.C.: USAF, 2002).

Essay on Sources

Little secondary historical literature directly bears on the subject of systems management or its precursors, although there is a wealth of literature about the Cold War, economic development, technology transfer, sociological communities, the aerospace industry, and so on. There are many references on the major organizations and politics of the time, but there is almost no prior historical research on the particulars of this topic. The scholarly inspiration for my work came mostly from the work of JoAnne Yates, *Control Through Communication: The Rise of System in American Management* (Baltimore: Johns Hopkins University Press, 1989) and James R. Beniger, *The Control Revolution: Technological and Economic Origins of the Information Society* (Cambridge: Harvard University Press, 1986) on the criticality of communication for managerial control. Howard McCurdy, *Inside NASA* (Baltimore: Johns Hopkins University Press, 1993) touches on some of these issues for the National Aeronautics and Space Administration (NASA) manned program, as does Thomas P. Hughes, *Rescuing Prometheus* (New York: Pantheon Press, 1998) on Atlas and the spread of the systems approach.

The history of management gained attention with the publication of Alfred D. Chandler Jr., *The Visible Hand: The Managerial Revolution in American Business* (Cambridge: Belknap Press of the Harvard University Press, 1977). Daniel Nelson's *Frederick W. Taylor and the Rise of Scientific Management* (Madison: University of Wisconsin Press, 1980) and Daniel Nelson, ed., *A Mental Revolution: Scientific Management since Taylor* (Columbus, Ohio: Ohio State University Press, 1992) detail the origin and spread of scientific management, as does Robert Kanigel's *The One Best Way, Frederick Winslow Taylor and the Enigma of Efficiency* (New York: Penguin Putnam, 1998). JoAnne Yates's work is the basic text for systematic management. Stephen P. Waring, *Taylorism Transformed: Scientific Management Theory since 1945* (Chapel Hill, N.C.: University of North Carolina Press, 1991) is an excellent work, albeit from a neo-Marxist viewpoint.

Literature on the history of industrial research is typified by Leonard Reich, *The Making of American Industrial Research: Science and Business at GE and Bell, 1876–1926* (Cambridge: Cambridge University Press, 1980) or the more recent work by David Hounshell and Johns Kenly Smith Jr., *Science and Corporate Strategy: Dupont R&D 1902–1980* (Cambridge: Cambridge University Press, 1988). William Minoru Tsutsui, *Manufacturing Ideology: Scientific Management in Twentieth-Century Japan* (Princeton, N.J.: Princeton University Press, 1998) is a good complement to this study in terms of its analysis of management in Japanese manufacturing. Charles Perrow's pathbreaking *Normal Accidents* (New York: Basic Books, 1984) addresses issues of reliability in complex systems in a way similar to this study.

Most of the material for chapters 2 and 3 comes from the Air Force Historical Research Agency at Maxwell Air Force Base in Montgomery, Alabama. The most useful collections there are the papers of Marvin Demler and Bernard Schriever, along with the Air Force Command Histories of Air Research and Development Command, Air Material Command, and Air Force Systems Command. The most valuable secondary sources were Irving Brinton Holley Jr., *Buying Aircraft: Materiel Procurement for the Army Air Forces,* vol. 7 of Stetson Conn, ed., *United States Army in World War II* (Washington, D.C.: Dept. of the Army, 1964); John Lonnquest's recent thesis on Schriever, "The Face of Atlas: General Bernard Schriever and the Development of the Atlas Intercontinental Ballistic Missile 1953–1960" (Ph.D. diss., Duke University, 1996); Davis Dyer, *TRW: Pioneering Technology and Innovation since 1900* (Boston: Harvard Business School Press, 1998); Jacob Neufeld, *Ballistic Missiles in the United States Air Force 1945–1960* (Washington, D.C.: Office of Air Force History, United States Air Force, 1990); and Robert Perry, *System Development Strategies: A Comparative Study of Doctrine, Technology, and Organization in the USAF Ballistic and Cruise Missile Programs, 1950–1960,* RAND Corporation Memorandum RM-4853-PR (Santa Monica, Calif.: RAND Corporation, August 1966). The Samuel Phillips Papers at the Library of Congress were also useful for the McNamara-Schriever interactions that created phased planning. Michael H. Gorn's *Vulcan's Forge: The Making of an Air Force Command for Weapons Acquisition (1950–1985)* (Washington, D.C.: Office of History, Air Force Systems Command, 1989) was an excellent source of information, as were command histories by Ethel M. DeHaven: *Air Materiel Command Participation in the Air Force Missiles Program through December 1957* (Inglewood, Calif.: Office of Information Services, Air Materiel Command Ballistic Missile Office, 1958) and *Aerospace: The Evolution of USAF Weapons Acquisition Policy 1945–1961* (Los Angeles: Deputy Commander for Aerospace Systems Historical Office, August 1962). So too is Ivan A. Getting's autobiography, *All in a Lifetime: Science in the Defense of Democracy* (New York: Vantage, 1989). The *AFSC Management Conference,* held at Monterey, California, 2–4 May 1962—published in Washington, D.C., by Air Force Systems Command—contained a number of very informative papers delivered by air force officers, scholars, and industry leaders. Critical information about configuration management's origins can be found in Bellis's paper from the conference: Benjamin N. Bellis, "The Requirements for Configuration Management During Concurrency," in *AFSC Management Conference,* Monterey, Calif., 2–4 May 1962 (Washington, D.C.: Air Force Systems Command, 1962). In the 1950s, a number of articles in the trade journal *Aviation Age* were insightful for deciphering the implications of changes in air force procurement methods.

Chapter 4 describes the development of systems management at the Jet Propulsion

Laboratory. Two secondary sources were extremely helpful, Clayton Koppes, *JPL and the American Space Program: A History of the Jet Propulsion Laboratory* (New Haven: Yale University Press, 1982) and R. Cargill Hall's *Lunar Impact: A History of Project Ranger,* SP-4210 (Washington, D.C.: NASA, 1977). Materials for this section came from the JPL Historical Archives in Pasadena, California. JPL materials were detailed and varied, with internal correspondence and reports to supplement those submitted to the army and NASA. JPL published very detailed final reports of its early projects that are excellent historical sources. Ranger internal and congressional investigations provided first-rate materials. An interview with Jack James was very valuable, clarifying some of the motivations behind JPL's politics and the "progressive design freeze."

Chapter 5 on the manned program has by far the most secondary sources from a number of NASA institutional and project histories, as well as other historical works by participants, journalists, and historians. NASA has two early administrative histories, one by Robert Rosholt, *An Administrative History of NASA, 1958–1963,* SP-4101 (Washington, D.C.: NASA, 1966) and Arnold S. Levine, *Managing NASA in the Apollo Era,* SP-4102 (Washington, D.C.: NASA, 1982). Of these two, Levine's is the more useful. However, their headquarters point of view turns out to be only marginally useful to understanding the dynamics of project management and systems engineering. NASA project histories were more useful—Mercury, Gemini, Saturn V, and Apollo—as were the NASA Project Chronologies (compilations and abstracts of the primary sources used by NASA historians to write their later project histories). The chronologies contain listings, abstracts, and lengthy reprints of the content of the major documents and are "semi-primary" sources. Also, the recent history of MSFC by Andrew J. Dunar and Stephen P. Waring, *Power to Explore: A History of Marshall Space Flight Center, 1960–1990,* SP-4313 (Washington, D.C.: NASA, 1999) is an excellent source for the organizational culture of Marshall. W. Henry Lambright's *Powering Apollo: James E. Webb of NASA* (Baltimore: Johns Hopkins University Press, 1995) has useful information about the role of NASA Administrator James Webb. Mike Gray's *Angle of Attack: Harrison Storms and the Race to the Moon* (New York: Penguin, 1992) was surprisingly insightful for a view from the contractor North American, as was William B. Harwood, *Raise Heaven and Earth: The Story of Martin Marietta People and Their Pioneering Accomplishments* (New York: Simon and Schuster, 1993) for Martin Marietta. Phillip K. Tompkins, *Organizational Communication Imperatives: Lessons of the Space Program* (Los Angeles: Roxbury, 1992) is a relatively little-known book that provides significant information about Wernher von Braun's organizational methods at Marshall Space Flight Center.

For this chapter I used primary sources from the Samuel Phillips Papers at the Library of Congress, the NASA Headquarters History Office, the Jet Propulsion Laboratory Archives, and congressional investigations and reports. Another very useful source is *Apollo Program Management,* Staff Study for the Subcommittee on NASA Oversight of the Committee on Science and Astronautics, U.S. House of Representatives, 91st Congress, 1st Session, July 1969 (Washington, D.C.: Government Printing

Office, 1967). This document is remarkable in that it lets NASA and its contractors showcase their management methods. Because Apollo was considered a great success at this time, NASA did not have to hide problems, so it was remarkably candid.

Chapters 6 and 7 describe the institutional structure and history of the European Space Vehicle Launcher Development Organisation (ELDO), the European Space Research Organisation (ESRO), and the European Space Agency (ESA). Three books discuss European space programs: Guy Collins's *Europe in Space* (New York: St. Martin's Press, 1990); John Krige and Arturo Russo, *Europe in Space 1960–1973,* SP-1172 (Noordwijk, The Netherlands: ESA, 1994); and the two-volume set by John Krige and Arturo Russo, *A History of the European Space Agency, 1958–1987,* SP-1235 (Noordwijk, The Netherlands: ESA, 2000). Douglas R. Lord, *Spacelab: An International Success Story,* SP-487 (Washington, D.C.: NASA, 1987) gives a very detailed story about that critical project. ELDO documentation is sparse, because many records disappeared when the organization shut down. However, an investigation into the failure of *Europa II* provides an unsparing appraisal of ELDO's problems: *European Space Vehicle Launcher Development Organisation, Europa II Project Review Commission, Final Report,* Neuilly, France, 18 April 1972, ELDO/CRP(72)40. ELDO, ESRO, and early ESA papers reside at the Historical Archives of the European University Institute in Florence, Italy. These are mostly official institutional papers, making it difficult to isolate individuals. ESRO and ESA records are relatively complete for top-level institutional decisions but are much less complete for lower-level technical divisions and directorates. The *ESA Bulletin* and its precursors in ESRO and ELDO provide much useful information, with articles aimed at a general audience. In the early 1970s, ESRO held a summer school to train its managers, leading to a summary of methods and projects documented in ESRO, *Project Management and Project Control, 10th ESRO Summer School,* Frascati, Italy, September 1972. There is also a substantial collection at the Technical Information and Documents Centre at the European Space Technology Centre (ESTEC), Noordwijk, The Netherlands.

For all chapters, the best sources were memoranda and letters describing the reasons for the adoption of new management and technical processes. These are rare and difficult to locate. The Samuel Phillips and Marvin Demler papers are conspicuous exceptions, with rich sources of this kind. I found it useful to collect project schedules, status reports, and weekly meeting minutes, because these sources indirectly describe processes; problems and events are unearthed through processes that these sources occasionally mention or describe.

Government investigations were extremely useful when they existed, such as *Final Report of the Ranger 6 Review Board* (Washington, D.C.: NASA, 17 March 1964) and U.S. Senate Report 956, Senate Committee on Aeronautical and Space Sciences, *Apollo 204 Accident Report,* 90th Congress, 2nd Session, 30 January 1968, and the final flight of *Europa II.* These give candid views of management practices and processes that are otherwise almost impossible to find. Lastly, there are a few articles written by managers and engineers regarding the management and systems engineering on their projects, usually given at professional meetings, for Congress, or for the Europeans, in the *ESA Bulletin.* Because these were not "sales" items like proposals and companies did not consider organizational processes proprietary, managers and engineers generally gave candid descriptions and assessments. I corroborated their accounts when I had internal data.

Three 1960s books on aerospace management were useful for this study: Fremont Kast and James Rosenzweig, eds., *Science, Technology, and Management* (New York: McGraw-Hill, 1963); Merton J. Peck and Frederic M. Scherer, *The Weapons Acquisition Process: An Economic Analysis* (Cambridge: Division of Research, Graduate School of Business Administration, Harvard University, 1962); and James E. Webb, *Space Age Management* (New York: McGraw-Hill, 1969).

Interviews were also a fruitful source of information. Historians' interviews some years ago with Donald Putt, George Mueller, and Samuel Phillips were helpful. My interviews with Bernard Schriever, Jack James, Ivan Getting, Simon Ramo, and Charles Terhune were extremely informative.

I have written a number of related articles that are published or are in the process of publication in journals such as *Technology and Culture, History and Technology, History of Technology, Airpower History,* and *Quest: The History of Spaceflight Quarterly.* Particularly important information on the air force can be found in *The United States Air Force and the Culture of Innovation 1945–1965* (Washington, D.C.: USAF History Support Office, 2002).

Primary sources were acquired from a number of locations and organizations:

- Air Force Historical Research Agency, Maxwell Air Force Base, Montgomery, Ala. (AFHRA)
- Charles Babbage Institute Archives, University of Minnesota, Minneapolis, Minn. (CBI)
- Technical Information and Documents Centre, European Space Agency, European Space Technology Centre, Noordwijk, the Netherlands (ESTEC)
- Historical Archives of the European University Institute, Florence, Italy (HAEUI)
- Archives and Records Office, Jet Propulsion Laboratory, Pasadena, Calif. (JPLA)
- Library of Congress, Samuel Phillips Papers, Washington, D.C. (LC/SPP)
- History Office, Headquarters, National Aeronautics and Space Administration, Washington, D.C. (NASAHO)
- USAF Space and Missile Center, El Segundo, Calif. (SMC)

Most information from these sources is readily accessible through finding aids. This is unfortunately not true for the Technical Information and Documents Centre (TIDC) at ESTEC. While its official documentation (the Engineering Working Papers [EWPs] and other formal documents) can be accessed through its computer system, most of the records from ESTEC are located in an unmarked file cabinet. Because of the small size of this cabinet, it is not unduly difficult to locate the documents I reference. TIDC personnel are very helpful and will show these files if asked. The NASA headquarters references suffer from the fact that on my first trip there I was not consistent with notating the original source boxes and folders. Some of the materials from this location are only marked as from the History Office, without further information. I have copies of all of these materials.

Index

Page numbers in *italics* denote illustrations.